IFIP Advances in Information and Communication Technology

518

Editor-in-Chief

Kai Rannenberg, Goethe University Frankfurt, Germany

Editorial Board

IFIP – The International Federation for Information Processing

IFIP was founded in 1960 under the auspices of UNESCO, following the first World Computer Congress held in Paris the previous year. A federation for societies working in information processing, IFIP's aim is two-fold: to support information processing in the countries of its members and to encourage technology transfer to developing nations. As its mission statement clearly states:

> IFIP is the global non-profit federation of societies of ICT professionals that aims at achieving a worldwide professional and socially responsible development and application of information and communication technologies.

IFIP is a non-profit-making organization, run almost solely by 2500 volunteers. It operates through a number of technical committees and working groups, which organize events and publications. IFIP's events range from large international open conferences to working conferences and local seminars.

The flagship event is the IFIP World Computer Congress, at which both invited and contributed papers are presented. Contributed papers are rigorously refereed and the rejection rate is high.

As with the Congress, participation in the open conferences is open to all and papers may be invited or submitted. Again, submitted papers are stringently refereed.

The working conferences are structured differently. They are usually run by a working group and attendance is generally smaller and occasionally by invitation only. Their purpose is to create an atmosphere conducive to innovation and development. Refereeing is also rigorous and papers are subjected to extensive group discussion.

Publications arising from IFIP events vary. The papers presented at the IFIP World Computer Congress and at open conferences are published as conference proceedings, while the results of the working conferences are often published as collections of selected and edited papers.

IFIP distinguishes three types of institutional membership: Country Representative Members, Members at Large, and Associate Members. The type of organization that can apply for membership is a wide variety and includes national or international societies of individual computer scientists/ICT professionals, associations or federations of such societies, government institutions/government related organizations, national or international research institutes or consortia, universities, academies of sciences, companies, national or international associations or federations of companies.

More information about this series at http://www.springer.com/series/6102

Eunika Mercier-Laurent · Danielle Boulanger (Eds.)

Artificial Intelligence for Knowledge Management

4th IFIP WG 12.6 International Workshop, AI4KM 2016
Held at IJCAI 2016
New York, NY, USA, July 9, 2016
Revised Selected Papers

 Springer

Editors
Eunika Mercier-Laurent ⓘ
Jean Moulin University Lyon 3
Lyon
France

Danielle Boulanger
Jean Moulin University Lyon 3
Lyon
France

ISSN 1868-4238   ISSN 1868-422X (electronic)
IFIP Advances in Information and Communication Technology
ISBN 978-3-030-06549-2   ISBN 978-3-319-92928-6 (eBook)
https://doi.org/10.1007/978-3-319-92928-6

Printed on acid-free paper

This Springer imprint is published by the registered company Springer International Publishing AG
part of Springer Nature
The registered company address is: Gewerbestrasse 11, 6330 Cham, Switzerland

Preface

The third wave of Artificial intelligence (AI) focuses more on exploring data than on knowledge. Most of the papers presented at the International Joint Conference on Artificial Intelligence (IJCAI) 2016 (http://www.ijcai-16.org/) were devoted to exploring exponentially growing data.

Knowledge management, which is still a challenge for many organizations, needs all facets of AI. This book aims to challenge researchers and practitioners in better exploring the three generations of AI and integrating international feedback from experience.

Knowledge management (KM) is a large multidisciplinary field with roots in management and AI. AI contributes to the way of thinking, knowledge modeling, knowledge processing, and problem-solving techniques. Knowledge is one of the intangible capitals that influence the performance of organizations and their capacity to innovate. Since the beginning of the KM movement in the early 1990s, companies and nonprofit organizations have experimented with various approaches.

Following the first AI4KM (Artificial Intelligence for Knowledge Management) organized by IFIP (International Federation for Information Processing) group TC12.6 (Knowledge Management) in partnership with ECAI (European Conference on Artificial Intelligence) held in 2012, and the second workshop hold during the Federated Conferences on Computer Science and Information Systems (Fedcsis) in 2014 in conjunction with the Knowledge Acquisition and Management Conference (KAM), the third manifestation initiated a partnership with IJCAI (International Joint Conference on Artificial Intelligence) in 2015. The Fourth AI4KM was held during IJCAI 16, in New York.

The objective of this multidisciplinary cooperation is still to raise the interest of AI researchers and practitioners in KM challenges, to discuss methodological, technical, and organizational aspects of AI used for KM, and to share feedback on KM applications using AI.

We would like to thank the members of the Program Committee, who reviewed the papers and helped put together an interesting program in New York. We would also like to thank all the authors. Finally, our thanks go to the local Organizing Committee and all the supporting institutions and organizations.

This volume contains selected papers presented during the workshop. After the presentation, the authors were asked to extend their proposals by highlighting their original thoughts. The selection focused on new contributions in any research area concerning the use of all AI fields for KM. An extended Program Committee evaluated the last versions of the proposals, leading to these proceedings.

We had an opportunity to host Dr. Janusz Wojtusiak from the Health Informatics Program, George Mason University, Fairfax on "Guiding Supervised Learning by Bio-Ontologies in Medical Data Analysis." His team paper describes the use of data semantics and ontologies in health and medical applications of supervised learning, and particularly describes how the Unified Medical Language System collaborate within the AQ21 rule-learning software.

This is followed by "Using Ontologies to Access Complex Data: Applications in Bio-Imaging." The authors propose a methodology combining a semantic approach with data management for more accurate searching and sharing of complex data in an organization.

The authors of "Dynamic Ontology Supporting Local Government" present their dynamic ontology-based knowledge model, designed to share and manage urban knowledge among representative of local self-government.

The next article entitled "Conceptual Navigation for Polyadic Formal Concept Analysis" describes the authors' formal concept analysis, which is a mathematically inspired knowledge representation with wide applications in knowledge discovery and decision support.

It is followed by "Highlighting Trendsetters in Educational Platforms by Means of Formal Concept Analysis and Answer Set Programming" presenting a Web-based educational system based on the analysis of the students' behavior aimed at developing methods to improve blended and e-learning systems.

The authors of "Selection of Free Software Useful in Business Intelligence: Teaching Methodology Perspective" propose methodology and a list of criteria for selecting free software tools to teach business intelligence especially adapted to small businesses.

The contribution entitled "Internet Platform for City Dwellers Based on Open Source Systems" presents the increasing intensification of activities of urban systems and proposes an Internet platform that will synchronize all processes occurring in the smart city and support the development of new ideas for improving life in the city.

This article is followed by "Segmentation of Social Network Users in Turkey," which presents an analysis of Turkish Twitter content using the self-organizing maps method. The results obtained demonstrate that by using segmentation, an important knowledge source can be derived and used as a tool for analyzing the market penetration of advertisements.

Finally, "Toward Semantic Reasoning in Knowledge Management Systems" examines the requirements and limitations of current commercial KM systems and proposes a new approach to semantic reasoning supporting big data access, analytics, reporting, and automation-related tasks. The resulting semantic-based analytics workflow was implemented for Siemens power generation plants.

The papers in this volume cover topics at the intersection of machine learning, knowledge models, KM and Web, knowledge capturing and learning, and KM and AI.

We hope you will enjoy reading these papers.

January 2018

Eunika Mercier-Laurent
Danielle Boulanger
Mieczyslaw Owoc

Organization

Co-editors

Eunika Mercier-Laurent   Jean Moulin University Lyon 3, France
Danielle Boulanger   Jean Moulin University Lyon 3, France
Mieczyslaw L. Owoc   University of Economics, Wroclaw, Poland
          (co-editor of e-proceedings from workshop
          http://ifipgroup.com/)

Program Committee

Danielle Boulanger   Jean Moulin University Lyon 3, France
Eunika Mercier-Laurent   Jean Moulin University Lyon 3, France
Nada Matta   Troyes Technical University, France
Mieczysław Lech Owoc   Wroclaw University of Economics, Poland
Otthein Herzog   Jacobs University, Bremen, Germany
Daniel O'Leary   USC Marshall School of Business
Antoni Ligęza   AGH University of Science and Technology, Poland
Helena Lindskog   Linköping University, Sweden
Gülgün Kayakutlu   Istanbul Technical University, Turkey
Knut Hinkelmann   University of Applied Sciences and Arts, Switzerland
Frédérique Segond   Viseo Innovation, France
Guillermo Simari   Universidad Nacional del Sur in Bahia Blanca,
          Argentina
Janusz Wojtusiak   George Mason University, USA

Local Organizing Committee

IJCAI-16

Eunika Mercier-Laurent   Jean Moulin University Lyon 3, France

Contents

Guiding Supervised Learning
by Bio-Ontologies in Medical Data Analysis

Janusz Wojtusiak[(✉)], Hua Min, Eman Elashkar, and Hedyeh Mobahi

Health Informatics Program, George Mason University, Fairfax, VA, USA
{jwojtusi, hmin3, eelashka, hmobahi2}@gmu.edu

Abstract. Ontologies are popular way of representing knowledge and semantics of data in medical and health fields. Surprisingly, few machine learning methods allow for encoding semantics of data and even fewer allow for using ontologies to guide learning process. This paper discusses the use of data semantics and ontologies in health and medical applications of supervised learning, and particularly describes how the Unified Medical Language System (UMLS) is used within AQ21 rule learning software. Presented concepts are illustrated using two applications based on distinctly different types of data and methodological issues.

Keywords: Supervised machine learning · Biomedical ontologies
UMLS

1 Introduction

Recent advancements of Machine Learning (ML) made it applicable to wide range of problems, including those in medical, healthcare and health domains. These methods are able to make accurate predictions in uncertain environments, by finding patterns scattered over massive amounts of data. Strength of many of the newest methods comes also in the ability to combine Natural Language Processing (NLP) tools with learning from structured data. The majority of novel methods are statistical and focus on analysis of numeric data. The principles of these machine leaning methods usually rely on distributions and patterns inside the data sets only. They do not include the meanings of the data sets. Domain knowledge for such methods is limited to ad-hoc encoding of attributes in the data or prior parameters of the model being learned as in examples of deep learning of neural networks [1]. Surprisingly, very few machine learning methods allow for modeling of domain knowledge in order to guide the learning process which can potentially make the learned models closer to human decision-makers.

The presented research explores the utilization of ontologies and data semantics to guide machine learning process. It provides the additional information to those distributions and patterns already inside the data sets. The motivations of this research come from the existence of rich biomedical ontologies created for the medical data integration methods, NLP algorithms and the Semantic Web [2], as well as early research on human-oriented machine learning. This presence of domain knowledge

E. Mercier-Laurent and D. Boulanger (Eds.): AI4KM 2016, IFIP AICT 518, pp. 1–18, 2018.
https://doi.org/10.1007/978-3-319-92928-6_1

provides an ideal opportunity to complement pure data for machine learning, with relationships that span over the data.

The researchers started to utilize domain knowledge to guide machine learning in some fields. For example, the frequent utilizations of the ontology exist in NLP, including ontology-based methods for indexing, extracting, and analyzing clinical notes [3–5]. In [6], the researchers classified patients with different types of epilepsy using different methodologies including ontology-based classification (OBC). The OBC achieved better results than others did.

After the brief introductions for biomedical ontologies and supervised machine learning, we will present an approach to handling semantic information and reasoning with ontologies in supervised learning.

1.1 Biomedical Ontologies

An ontology formally represents domain knowledge as a set of concepts and relationships between those concepts. The concepts in an ontology should be close to objects (physical or logical) in the real world. Relationships describe the interactions between concepts or a concept's properties. The most important relationship is the "IS-A" hierarchical relationship. It serves as the ontology's backbone and supports the property inheritance. The "IS-A" relationship connects a more specific concept (child concept) to a more general concept (a parent). Non-IS-A relationships, called associative or semantic relationships, connect concepts across the hierarchies in an ontology. Broadly speaking, ontologies include thesauri, terminologies, classifications, and coding systems. Ontologies play important roles in biomedical research through a variety of applications including data integration, knowledge management, natural language processing, and decision support [7].

The most popular biomedical ontologies (Bio-Ontologies) include Systematized Nomenclature of Medicine—Clinical Terms (SNOMED CT) [8], International Classification of Diseases (ICD) [9], Logical Observation Identifiers Names and Codes (LOINC) [10], Gene Ontology (GO) [11], Medical Subject Headings (MeSH) [12], RxNorm [13], Foundational Model of Anatomy (FMA) [14], and National Cancer Institute Thesaurus (NCI Thesaurus) [15]. Each ontology has its own purpose and scope. For example, SNOMED CT is a systematically organized computer processable ontology of medical terms. ICD defines the universe of diseases, disorders, injuries and other related health conditions. LOINC is a coding system for laboratory and clinical observations. GO provides controlled vocabularies of defined terms representing gene product properties including cellular components, molecular functions and biological processes. MeSH is designed to provide a hierarchically-organized terminology for indexing and cataloging of biomedical information such as MEDLINE/PubMed. RxNorm, published by National Library of Medicine (NLM), provides normalized names and a model for clinical drugs available in the US. FMA represents a coherent body of explicit declarative knowledge about human anatomy. Finally, NCI Thesaurus includes broad coverage related to the cancer research domain.

Therefore, there are communication barriers between various information systems or applications if the developers use different vocabularies in different systems. In order to solve these barriers, the Unified Medical Language System (UMLS) was developed by the NLM in 1986 [16] and it is constantly being updated. The UMLS has three knowledge sources: Metathesaurus, Semantic Network, and SPECIALIST Lexicon. The UMLS (2016AB) contains more than 3.44 million concepts (Concept Unique Identifiers; CUIs), 22 million relationships among those concepts, and 13.7 million unique concept names (AUIs) from 199 source vocabularies. One important goal of the UMLS is to establish mappings between different Bio-Ontologies. A concept unique identifier (CUI) is assigned to the terms from various source ontologies that have the same meaning in the Metathesaurus. The mappings among these vocabularies allow computer systems to translate data among the various information systems. The rich relationships in the UMLS also provide a solid foundation for reasoning in the medical knowledge [7]. Other UMLS applications include providing browser for its source ontologies, the Clinical Observations Recordings and Encoding (CORE) Problem List Subset [17], NLP [3–5], and value sets for Clinical Quality Measures (CQMs) [18].

1.2 Supervised Learning

While machine learning is a broad area, the presented work is focused on supervised learning, and more specifically concept learning. The methodology described here can be applied to output concepts which are independent, ordinal or structured. One can also extend the method presented here into regression learning, as well as other forms of machine learning, i.e. unsupervised and reinforcement learning.

The problem considered here is to learn a model $M: X \rightarrow Y$ which can be viewed as a function that assigns classes from $Y = \{y_j\}, j = 1..k$ into objects from X. Learning is performed by an algorithm A given dataset D and background knowledge BG, $A(D, BG) = M$, where $D = \{(x_i, y_i), i = 1..N\}$. This work focuses on the background knowledge, BG, and specifically its forms that can be retrieved from ontologies.

Many concept learning methods have been developed in the field, including symbolic methods for learning decision trees [19] or rules [20], numeric methods for learning sets of equations such as Support Vector Machines [21], Logistic Regression [22], or non-Negative Matrix Factorization [23]. Regardless of the used method, the general goal is to build models that maximize quality measure (or minimize loss function). An issue considered here is how to improve the methods when additional structure about the problem is known, i.e., in the form of hierarchical relationships between values of attributes, or non-IS-A relationships between attributes. In a sense, recent work on neural networks, related to deep learning [24] can be viewed as a form of encoding of problem into structure of model. In deep learning different structures of networks are considered, which can be based on hierarchies, but typically represent components of the problem rather than semantic concepts.

Early and more recent work on learning structures [25] and use of background knowledge included advanced in Inductive Logic Programming (ILP) and hierarchical learning methods. In the ILP, the background knowledge is defined as a set of relations (predicates) that can be used in the definition of the target concept [26]. An ILP system derives rules based on an encoding of the known background knowledge and a set of examples represented as a logical database of facts. Another related field is the hierarchical Relational Reinforcement Learning (HRRL) [27–29]. In those studies, hierarchies have been used to reduce the complexity of decision making and improve the actual process of learning. Both ILP and HRRL are particularly useful in bioinformatics, healthcare, and NLP.

1.3   Example Data

This paper illustrates concepts of using ontologies and semantics in machine learning on two examples of medical/health data. These are chosen because of inclusion of diverse types of data and are part of existing projects on using semantics in machine learning by the authors.

- **SEER-MHOS:** The example learning problem concerns the ability to automatically assess patient disabilities in performing Activities of Daily Living (ADLs). Such activities are important measures of patient independence, quality of life and need for care. However, the data about ADLs is not routinely collected along with clinical or administrative data. The purpose of this application is to automatically assess patients' functional disabilities based on general demographic information and known broad categories of diagnoses. Models are trained on SEER-MHOS (Surveillance, Epidemiology, and End Results – Medicare Health Outcomes Survey) which is a linked dataset. A subset of data with 1,849,311 unique patients, out which 102,269 patients diagnosed with cancer, have been used for this research. SEER is a cancer registry program that provides clinical, demographic and cause of death information [30]. MHOS data is a survey based report that contains both patient conditions and ADLs. In this research, SEER-MHOS data has been coded with UMLS CUIs and used to create models for predicting ADLs after cancer diagnosis.
- **MIMIC-III:** Learning from clinical data adds another level of complexity beyond standard administrative healthcare data. The MIMIC III ('Medical Information Mart for Intensive Care') is a large database that includes de-identified, comprehensive clinical data of patients admitted to critical care units at a large tertiary care hospital, Beth Israel Deaconess Medical Center in Boston, Massachusetts. The data are publicly available to researchers who satisfy certain conditions [31]. MIMIC-III consists of over 58,000 hospital admissions for 38,645 adults and 7,875 babies. It is structured into 26 tables organized as a relational database. Data have been collected during routine hospital care between 2001 and 2012 and was downloaded from several sources including archives from critical care information systems, hospital electronic health record and Social Security Administration Death Master File [32]. In the presented work, MIMIC-III data has been mapped to UMLS ontology and used to create models for predicting 30-day post-hospitalization mortality.

2   AQ21, Semantics, and Ontologies

AQ21 is the latest of rule learning systems developed by Ryszard Michalski's team at George Mason University and previously at University of Illinois [33]. Currently AQ21 is being extended by methods that allow for reasoning with complex data, i.e., data that is mapped into ontologies, data with multiple types, and specifics of medical and health data [34]. The AQ family of rule learners follow traditional separate-and-conquer approach to learning, by generating multiple stars (all rules one positive example, *the seed*, that do not cover negative examples). From each star top rules are selected for further processing. This operation is repeated until all positive examples are covered in the training data. Finally, the learned rules are optimized to maximize their quality according to user-defined criteria. AQ21 software implements several additional modules for adjusting representation space by constructive induction [35, 36], testing and applying data, handling time series, generating natural language descriptions from rules, and others. The research on AQ rule learners follows the idea of *natural induction* [37] in which created models are in forms natural to people (transparent and consistent with their prior knowledge).

The general principle behind the work presented here is that a rule learning system that understands semantics as well as relationships between attributes or values, can reason better than one that is provided only data. This principle is grounded on how people reason. Instead of purely relying on data, people use their knowledge to put all data in context and reason about it. For example, knowing that Type I Diabetes and Type II Diabetes are both Diabetes, reduces complexity and allows for the learned description to be more general (handling IS-A relationships and reasoning with hierarchies is described later in this chapter). By semantics of data, we understand attribute types, meta-values, aggregated vs. individual data, and relationships between data elements.

Attribute types are used to specify which methods of reasoning, including data transformations, can be applied when learning. The basic recognized attribute types are: nominal (unordered sets of symbolic values, i.e., *[treatment=radiation, surgery]*), ordinal (ordered sets of symbolic values, i.e., *[stage=I..III]*), cyclic (ordered sets of symbolic values that form a cycle, i.e., *[day=Friday..Monday]*), structured (hierarchical sets of symbolic values, i.e., *[treatment=surgery]* with surgery having subcategories such as robotic surgery, etc.), graph (values are linked together by edges in a graph, i.e.,), set (multiple values can be selected at the same time, i.e., *[diagnosis={diabetes, hypertension, obesity}]*), set-defined (similar to set, but with additional structure added on top of values, i.e., diagnoses in previous example form a hierarchy), interval (numeric values with defined addition and subtraction, zero is not defined), ratio (numeric values with defined multiplication and division), cyclic-ratio (numeric values forming a cycle, i.e., *[angle=276..15]*), and absolute (numeric values with only order defines and no operations permitted, i.e., social security number). These attribute types along with additional examples are explained in [38]. Attribute types are critical when attempting to generalize and reason with rules, as well as apply constructive induction methods, i.e. interval attributes should not be multiplied.

Aggregated data refer to data in which examples provided to the learning system describe a group of individual observations, rather than an individual object about which the system reasons [39]. Aggregated data are typically described using mean and standard deviation of attributes measured for a group of examples, or frequency of symbolic values. The learning problem from aggregated data is to create models for categorizing individual data when no individual training data are available, or only small portion of individual data are available with addition of aggregated data. Learning from aggregated data is particularly important in fields such as healthcare in which access to individual patient data is difficult or impossible. It is inspired by the field of meta-analysis in which aggregated data retrieved from published scientific studies can be analyzed to arrive at global conclusions supported by majority of studies.

Meta-values refer to special values present in the data, namely unknown, not applicable and irrelevant [40]. These meta-values correspond to potential reasons for which regular values are not present: they are not known, do not make sense, or are removed based on expert's judgment. The majority of machine learning and data mining methods ignore the fact that not all missing data are the same. Imputation methods can be meaningfully applied only to values that exist but are not known. Imputing data for not-applicable values simply does not make sense (for example replacing missing Prostate-specific Antigen (PSA) missing value for female patients in medical dataset). Similarly, imputing irrelevant values that were deliberately removed by experts makes no sense. In statistics, there is a distinction between data missing at random, missing completely at random, and missing not at random. While these classes partially correspond to semantic meaning of meta-values their interpretation is different. Meta-values represent background knowledge that is handled internally within learning systems such as AQ21.

Medical data in the electronic health record systems (EHRs) include a wide range of data from patient demographic, medical history, diagnosis, treatments, socioeconomic status, to genetic information. The medical data can be coded with concepts from one or multiple medical ontologies. Those concepts are connected by different type of relationships including most typically used IS-A relationships being part of hierarchies, or non-IS-A relationships that carry semantic meaning of the connections between concepts. The concepts and their relationships are represented in the medical ontologies in formal knowledge representation languages such as OWL and OBO. Thus, they are computationally processable by machine learning methods. In summary, semantic data description includes information about attribute types, inter-attribute relationships, value aggregation semantics, data transformations, and meta-values.

The main focus of the presented work is to use the semantics of data when applying supervised machine learning methods to construct classification models. The following sections describe how relationships can be extracted from UMLS and included as part of background knowledge used by AQ21. Generalization with hierarchies and IS-A relationships, learning with non-IS-A relationships, and finally learning hierarchical and ordinal models by AQ21 are presented.

2.1   Hierarchy Extraction from UMLS

The data pre-processing includes one major step that is a multi-step hierarchy extraction. It contains 4 steps: (1) Mapping data to the UMLS concepts (and identify their CUIs); (2) Extracting complete sub-hierarchy by following IS-A relationships using CUIs from step 1; (3) Resolving inconsistencies in the hierarchy (e.g., cycles, duplicates); and (4) Encoding extracted hierarchies in ML-software readable format (i.e., AQ21 requires a list of parent-child pairs for all relationships that form hierarchy). The detailed description is outlined below:

1. Map data to the UMLS concepts. The mapping is a challenge since the meaning of the concepts varies depending on specific sources and authors. For example, the

definition of Congestive Heart Failure (CHF) from Elixhauser et al. in 1998 [41] is different from UMLS. According to the definition from Elixhauser, there are 18 ICD-9 codes that are classified as CHF (see first column in Table 1). While in UMLS, to understand how CHF is defined, we need to follow hierarchy starting from the most general Congestive Heart Failure (CHF) concept, tracking down all its possible children through the hierarchy. This was done by tracking children of the concept, then children of all children, to last concept related to the main general concept, as shown in Fig. 1. However, there are only 11 concepts (CUIs) that defines Congestive Heart Failure (CHF) in the UMLS system (and corresponding ICD-9 codes) as shown in second column in Table 1. In this case, CUIs resulting from tracking hierarchy of CHF concept in UMLS, were mapped again to corresponding ICD9 so we can show concepts in both Elixhauser and UMLS defined using ranges of ICD9. The seven inconsistent codes are highlighted in gray in Table 1. This is just one example of the fact that the mapping process is difficult, and currently needs to be done manually by domain experts.

2. Follow IS-A relationships in the UMLS for both parents and children until complete sub-hierarchy is extracted. The hierarchy should extend to the farthest common ancestors and descendants. In UMLS one should follow relationships corresponding to all considered source terminologies, not only one in which the original data is coded.

3. Resolve inconsistencies in the extracted hierarchies. Due to nature of UMLS and fact that it is constructed from multiple source terminologies, a number of inconsistencies may happen. For example, the extracted hierarchies often include cycles, which are not permitted in AQ21 (and make it impossible to reason with data). Figure 2 shows one example of cycles in UMLS. Cycles are resolved by breaking links that go back to concepts higher in the hierarchy, as measured by distance from the UMLS root. Other types of inconsistencies include use of duplicate concepts or depreciated concepts. Those inconsistencies should be removed from the final hierarchy.

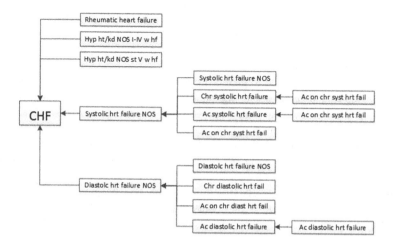

Fig. 1. Congestive Heart Failure concept hierarchy extracted from UMLS.

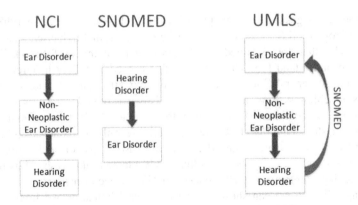

Fig. 2. Example cycle in UMLS created when combining multiple terminologies (SEER-MHOS Data).

4. Encode hierarchy in ML-software readable format, typically a list of parent-child pairs. AQ21 is a stand-alone software that reads input from text files. The files need to include all semantic information required to correctly reason with the data. Specifically, in AQ21 hierarchical relationships are part of definition of domains attributes that describe data.

Currently the steps 1–3 are completed outside of AQ21 software, but the system is being extended by the ability to link directly to UMLS. In principle, with right data pre-processing and representation transformations, it may be possible to encode semantic information in a way that most ML methods can use it.

2.2   Generalization with Hierarchies

Existing methods in AQ21 allow for using hierarchies extracted from ontologies to be used in specifying optimal generalization level of rules. The method is based on an approach first described by Kaufman and Michalski [42] and implemented in earlier AQ systems. The method follows IS-A relationships in the data to generalize or specialize rules in order to improve their accuracy and simplicity as defined by implemented rule quality measures.

The method can be applied to generalize beyond a single attribute in the data that implements set-valued attributes using binary indicators. This is particularly useful when analyzing coded medical data with potentially hundreds of thousands of binary attributes. For example, patient diagnoses coded using ICD-9 codes can result in the need to create close to 10,000 binary attributes. It is important to generalize ICD-9 codes in order to reduce the number of features. The generalization can be done by categorizing ICD-9 codes into Clinical Classifications Software (CCS) codes or finding a common ancestor for those codes by climbing the UMLS hierarchy. In this study, CCS was applied to generalize ICD-9 codes for the SEER-MHOS dataset. The goal was to find the predictor or set of predictors for the ADLs deficiencies.

Table 1. Congestive Heart Failure (CHF) as defined by Elixhauser et al. and in the UMLS hierarchy (MIMIC III Data).

Elixhauser Definition (ICD-9)	UMLS Definition (CUI)	Description
398.91	C0155582	Congestive rheumatic heart failure
402.01	Not CHF	Malignant hypertensive heart disease W heart failure
402.11	Not CHF	Benign hypertensive heart disease W heart failure
402.91	Not CHF	Unspecified hypertensive heart disease W heart failure
404.01	Not CHF	Hypertensive heart & chronic kidney disease, malignant, w heart failure & chronic kidney disease stage I - stage IV, or unspecified
404.03	Not CHF	Hypertensive heart & chronic kidney disease, malignant, w heart failure & chronic kidney disease stage V - end stage renal disease
404.11	Not CHF	Benign hypertensive heart & renal disease W CHF
404.13	Not CHF	Benign hypertensive heart & renal disease W CHF & renal failure
404.91	C3665458	Hypertensive heart & renal disease W heart failure & renal failure
404.93	C0494576	Heart Failure, Systolic
428.20	C1135191	Acute systolic heart failure
428.21	C2732748	Chronic systolic heart failure
428.22	C1135194	Acute on chronic systolic heart failure
428.23	C2733492	Heart Failure, Diastolic
428.30	C1135196	Acute diastolic heart failure
428.31	C2732951	Chronic diastolic heart failure
428.32	C2711480	Acute on chronic diastolic heart failure
428.33	C2732749	Hypertensive heart & renal disease W heart failure & renal failure

Before analysis, the data were preprocessed as follows: First we limited our study population to those patients who completed at least one survey before their cancer diagnosis and one survey roughly one year after the diagnosis. If a patient had multiple surveys, we used the surveys closest to before and after the cancer diagnosis. These very strict criteria significantly reduced the data size and the process produced a cohort of 723 cancer patients. The set of output attributes included six ADL indicators (walking, dressing, bathing, moving in/out chair, toileting, and eating) that were extracted from the survey completed after the cancer diagnosis. Input attributes, extracted from survey completed prior to cancer diagnosis and cancer registry, were based on known ADL factors from the literature [43–46]. They include patient demographic (age, race, marital stats), ADLs before cancer diagnoses, comorbidities (Diabetes, Hypertension, Arthritis, etc.), cancer characteristics (tumor size, staging, etc.), surgery and treatment indicators. The ICD-9 codes in the final dataset were mapped to the UMLS CUIs. These CUIs were used to extract the hierarchical relationships (Parents and Children) from UMLS until a common ancestor was found. The extracted hierarchies were added to the set of input variables.

AQ21 software was used to investigate the method with and without using background knowledge from UMLS. Application of the AQ21 software to SEER-MHOS data mapped to UMLS resulted in a number of models (rulesets) for predicting patients' deficiencies in performing activities of daily living. AQ21 has been executed in Approximate Theory Formation Mode (ATF), with weight w = 0.3 of completeness vs. consistency gain. In the ATF mode, AQ21 produces rules that may be partially incomplete or inconsistent in order to maximize the rule quality measure. Below are two sample rules generated by AQ21 with and without using background knowledge:

Sample 1: AQ21
[Bathing impairment] <== [Race = 2,1,4: 70, 245, 22%]
                         [Hispanic = 2: 64, 241, 20%]
                         [Smoking = 2,3: 68, 238, 22%]
                         [Surgery = 51,40,27,0,45: 45, 113, 28%]
                         [Histology = 2,4,5,15,8,9,1: 74, 252, 22%]
                         [Stage = 0,1,2: 69, 244, 22%]
                         [Primary site and morphology = C0153458,
                         C0153492, C0153532, C0242787, C0949022,
                         C0235653, C0153483, C0153611, C0153555,
                         C0153435, C0346782, C0153491, C0153612:
                         30, 34, 46%]
                         : p = 22, n = 2, q = 0.642

Sample 2: AQ21 with background knowledge
[Bathing impairment] <== [Race = 1,4: 64, 219, 22%]
                         [Hispanic = 2: 64, 241, 20%]
                         [Smoking = 2,3: 68, 238, 22%]
                         [Surgery = 32,51,40,0,45: 40, 95, 29%]
                         [Histology = 2,5,15,8,9,1: 68, 229, 22%]
                         [Cancer site = 2030, 25010, 21047, 21052,
                         21100, 29010,26000, 22020: 61, 169, 26%]
                         [Primary site and morphology = C0154077,
                         C0007102, C0153532, C0005684, C0153555,
                         C0024624, C0006142, C0235652, C0864875,
                         C0346647, C0345921, C0242379, C0346629,
                         C0345865, C0242788, C0034885, C0007107,
                         C0345713, C0587060, C1263771: 38, 49, 43%]
                         : p = 23, n = 2, q = 0.653

The rules are similar using AQ21 with and without background knowledge. However, the quality of rule, as measured by Q(w), generated by the second method is improved. The last line in the rule set describes the numbers of positive examples (p), negative example (n) covered by the rule, and the rule quality. While the rules presented above correspond to each other, AQ21 with and without ontology are not guaranteed to create similar rules. Instead the program applies beam search to go through space of possible combinations of attributes and values to find the highest quality rules. Presence of additional ways of generalizing data available in the presence of hierarchies derived from an ontology may steer the process in different direction. Consequently, the quality of rules improves because of the ability to generalize using hierarchies.

2.3   Using Non-IS-A Relationships

The current version of UMLS includes 727 types of relationships (NLM, 2016), with IS-A being just one of them. Semantics of these relationships need to be encoded in learning software, and their use and effect on reasoning process depends on specific meaning of that relationship. One simple way of encoding these relationships is that the data can then be extended by additional dimensions that correspond to presence of meaningful relationships. The following process is used to search for additional attributes to be added to problem representation space. It checks all pairs of attributes in the data and their values for existence of relationships. The following steps describe the method in terms of using UMLS, but can be easily extended to other ontologies.

1. Map the used attributes in the dataset (e.g., ICD-9) to the corresponding UMLS concept unique identifiers (CUIs). This can be done automatically if the coding system used to the attributes in data is part of UMLS' source vocabularies. Otherwise it can be done manually by experts.
2. For each pair of concepts retrieved:
   2.1. Search UMLS for non-hierarchical relationship(s) between the two concepts and all their parents (generalize using IS-A to the closest parents, or those within a predefined distance)
   2.2. Add all found relationships to a list of candidate attributes, and add new attributes to the data that indicates presence of the relationship. An example is shown in Figure 3. Concept X and Y exist in the dataset (see Figure 3a). The non-hierarchical relationship between X and Y is extracted from UMLS (see Figure 3b) and added as a new attribute X_Y to the dataset (see last column in Figure 3a). If the X and Y present in a patient's record, the value of the new attribute X_Y should be 1. Otherwise, it should set to 0.
   2.3. Apply attribute selection methods to filter out potentially large number of new relation-based attributes from the data. Those methods select attributes for learning by computing the discriminatory power of each attribute and comparing it with the acceptance threshold. Attributes whose discriminatory power is below the threshold will not be used for learning.
3. Apply standard learning algorithms on the data. At this point the data consists of original attributes including those mapped to UMLS and additional binary attributes that represent relationships. This encoding allows the use of any learning methods applicable to the original dataset.

The above method generates potentially very large number of new attributes, significantly increasing size of representation space for learning. The data are also typically sparse because of frequency of the related concepts co-occur in the data. Thus, efficient attribute selection methods (Step 2.3) need to be applied to reduce dimensionality. Experimental results performed on MIMIC III data indicate that even for large datasets there is a need to select most relevant attributes.

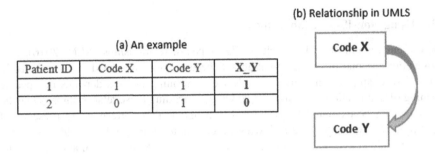

(a) An example

Patient ID	Code X	Code Y	X_Y
1	1	1	1
2	0	1	0

(b) Relationship in UMLS

Code X

Code Y

Fig. 3. Example candidate attribute

MIMC III contains rich clinical data including diagnoses and treatments for critical care patients. For example, some patients diagnosed with respiratory tract disease were treated by prednisone. As depicted in Fig. 4, the non-IS-A relationship between "respiratory tract disease" and "prednisone" were extracted from the UMLS. Prednisone is a drug that "may treat" respiratory tract disease. Not surprisingly, the ontology includes also the reverse relationship "may be treated by". According to the Step 2.2, a new relation-based attribute "respiratory tract disease - prednisone" was added to the data. The value of the new attribute should be "1" for those patients for whom prednisone was used to treat that condition.

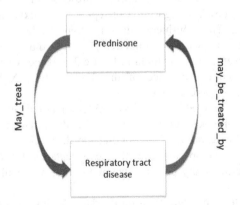

Fig. 4. Example non-IS-A relationship within MIMIC III data extracted from UMLS.

In order to test the described method for using non-IS-A relationships, it was applied to MIMIC III dataset in order to construct models for predicting 30-day post-hospitalization mortality. The performance of the two methods, with and without using semantic (non-IS-A relationships), were compared. The primary input attributes included the diagnoses registered during hospitalization. In order to prepare data, patients with age 65 and older and their admission records were extracted from the data (25,525 hospitalization records). Diagnoses originally coded with ICD-9 codes in the data were mapped to Clinical Classifications Software (CCS) codes [48] to reduce the number of features in the data and group conditions into clinically meaningful categories.

As illustrated in Fig. 3, candidate relationships were added as new attributes to the input set. CCS codes were mapped to CUIs. The IS-A relationships in UMLS is then followed to create the neighborhood space (parents and children) for each CUI. The neighborhood was used to find all non-IS-A relationships between each paired concepts. As a result, this method generated a relatively large number of new attributes (443) compared with the sample size. Hence, feature selection methods were used to decrease the size of representation space. The final dataset containing 421 features was used to learn models for predicting mortality.

Finally, standard learning algorithms such as Bayes network, naïve Bayes and logistic regression were applied to the dataset with and without using semantic. After adding semantics to the method, most models were able to capture more true positive cases and achieved higher recall which can be crucial in case of mortality prediction. For example, the naïve Bayes method without using semantics correctly classified 263 out of 555 death cases, while this method with semantics captures 10 more true positive instances.

2.4 Learning Ordinal and Hierarchical Outputs

A typical approach to building multiclass classifiers is to learn models for each class against all other classes in the data as shown in Fig. 5a. While effective in many cases, this approach suffers particularly when dealing with problems with many classes or when there is inherit structure to the concepts being described. Independent binary classifiers also do not allow for weighting types of mistakes made during classification (i.e., classification error of diabetes vs. cancer is worse than one of type I and type II diabetes).

Building hierarchical classifiers has recently become popular approach in machine learning applications. Instead of building classifiers with large number of unrelated classes, information about structure of output (Fig. 5c) may significantly improve performance of learning algorithms. Moreover, errors at lower levels of hierarchy are less critical than those at higher levels [42]. In order to build a hierarchical classifier, AQ21 starts with building models that distinguish general concepts at the top of the hierarchy (those connected to the root). Then data is limited to those within one general concept and models are built to describe sub-concepts. This operation is repeated recursively until all concepts in the hierarchy have corresponding models. AQ21 implements the method in breath-first search strategy but the order does not affect results nor computation time.

In addition to hierarchical structure within input attributes, AQ21 allows for ordered structures of output attributes. An example of ordinal (ordered) output is when one considers three or more levels of patient disability. A patient may be fully independent, partially dependent/needs some help, or fully dependent in performing a certain task. In order to learn ordinal output, the system will first build a model to distinguish between fully independent patients and those with any level of dependence, and then among the dependent patients, distinguish between those with partial and full dependence, as illustrated in Fig. 5b. It is clear that the order of values in the domain of ordinal attribute affects results of learning. It can be also easily observed that the order high-to-low will result is completely different classifier than low-to-high, thus one needs to carefully design attribute domains.

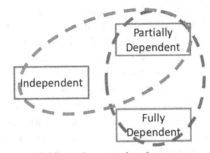

(a) Learning unordered output.

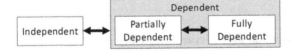

(b) Learning ordinal structure of output.

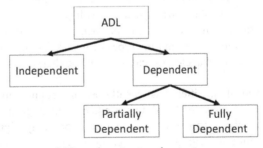

(c) Learning structured output.

Fig. 5. Learning unstructured and structured outputs.

3 Conclusions

Over the past decade significant progress has been made in the ability to use semantic information and ontologies in machine learning, despite being outside of mainstream research in the field. Research of our group at George Mason University focuses on selected aspects on using semantics (meta-values, ontologies, aggregated data) and is currently done in the context of analyzing medical, healthcare and health data.

In this study, we explored to include both hierarchical and non-hierarchical relationships in our data analysis. Both methods helped the learning process when models were created form the SEER-MHOS and MIMIC datasets. We have demonstrated that adding semantics to the ML method improved the performance of the prediction model by achieving higher recall. Capturing more true positive cases in the prediction model is important in some areas like predicting mortality or ICU admission. The results of the developed ML need to be interpreted by the domain experts. The new rules found by this study will be new hypotheses and validated by future investigations.

The presented preliminary study has a number of limitations. Several steps of data preparation need to be done manually, such as mapping data to the UMLS concepts. This causes problems related to resolving ambiguity between concepts and relationships among various ontologies, health agencies, users, and the way these ontologies are used. Different relationships between concepts may be important when a learning model preforms different tasks (i.e., differential diagnoses, comparative effectiveness of treatments, outcome prediction). Using UMLS in the ML is also challenge due to its extremely large size and high complexity. When we implemented the rich knowledge from UMLS and added them as new attributes, the dimension of the dataset increased dramatically. Thus, it brings another challenge for developing efficient attribute selection algorithms for data reduction. Currently our method increased only accuracy of models by small fraction. This may be the case that our simple strategy for hierarchical and non-hierarchical is not sophisticated enough to significantly improve the overall performance of our machine learning algorithm. Further research should be done to implement more sophisticated background knowledge and take greater advantage of the structure of the datasets. It is also important to study the effectiveness of problem of increasing dimensionality in context of size of data that is used for learning.

New wave of interest in artificial intelligence opens promise that methods using semantics and making computers "understand" objects which they reason or learn from will return to attention in machine learning. The most important work to be done in the near future concerns the ability to combine incredible advances make in statistical machine learning methods, with techniques described in this paper. High predictive accuracy of statistical models that also make sense to human experts is particularly important in domains such as healthcare where transparency is critical to users.

Acknowledgments. Our current activities on using semantic in machine learning are supported part by the Jeffers Foundation and LMI Academic Partnership Program.

References

1. Gülçehre, Ç., Bengio, Y.: Knowledge matters: importance of prior information for optimization. J. Mach. Learn. Res. **17**(8), 1–32 (2016)
2. Tresp, V., Bundschus, M., Rettinger, A., Huang, Y.: Towards machine learning on the semantic web. In: da Costa, P.C.G., d'Amato, C., Fanizzi, N., Laskey, K.B., Laskey, K.J., Lukasiewicz, T., Nickles, M., Pool, M. (eds.) URSW 2005-2007. LNCS (LNAI), vol. 5327, pp. 282–314. Springer, Heidelberg (2008). https://doi.org/10.1007/978-3-540-89765-1_17
3. Cai, T., Giannopoulos, A.A., Yu, S., et al.: Natural language processing technologies in radiology research and clinical applications. Radiographics **36**(1), 176–191 (2016)
4. Wu, S.T., Liu, H., Li, D., Tao, C., Musen, M.A., Chute, C.G., Shah, N.H.: Unified Medical Language System term occurrences in clinical notes: a large-scale corpus analysis. J. Am. Med. Inform. Assoc. JAMIA **19**(e1), e149–e156 (2012)
5. Xu, R., Musen, M.A., Shah, N.H.: A comprehensive analysis of five million UMLS metathesaurus terms using eighteen million MEDLINE citations. In: AMIA Annual Symposium Proceedings, pp. 907–911 (2010)
6. Kassahun, Y., et al.: Automatic classification of epilepsy types using ontology-based and genetic-based machine learning. Artif. Intell. Med. **61**(2), 79–88 (2014)
7. Bodenreider, O.: Biomedical ontologies in action: role in knowledge management, data integration and decision support. Yearb. Med. Inform. **47**, 67–79 (2008)
8. Stearns, M.Q., Price, C., Spackman, K.A., Wang, A.Y.: SNOMED clinical terms: overview of the development process and project status. In: Proceedings of the AMIA Symposium, pp. 662–666 (2001)
9. Hirsch, J.A., Nicola, G., McGinty, G., Liu, R.W., Barr, R.M., Chittle, M.D., Manchikanti, L.: ICD-10: history and context. AJNR Am. J. Neuroradiol. **37**(4), 596–599 (2016)
10. Huff, S.M., Rocha, R.A., McDonald, C.J., et al.: Development of the Logical Observation Identifier Names and Codes (LOINC) vocabulary. J. Am. Med. Inform. Assoc. JAMIA **5**(3), 276–292 (1998)
11. The Gene Ontology Consortium: Expansion of the Gene Ontology knowledgebase and resources. Nucleic Acids Res. **45**(Database issue), D331–D338 (2017)
12. Coletti, M.H., Bleich, H.L.: Medical subject headings used to search the biomedical literature. J. Am. Med. Inform. Assoc. JAMIA **8**(4), 317–323 (2001)
13. Nelson, S.J., Zeng, K., Kilbourne, J., Powell, T., Moore, R.: Normalized names for clinical drugs: RxNorm at 6 years. J. Am. Med. Inform. Assoc. JAMIA **18**(4), 441–448 (2011)
14. Rosse, C., Mejino Jr., J.L.: A reference ontology for biomedical informatics: the Foundational Model of Anatomy. J. Biomed. Inform. **36**(6), 478–500 (2003)
15. Fragoso, G., de Coronado, S., Haber, M., Hartel, F., Wright, L.: Overview and utilization of the NCI thesaurus. Comp. Funct. Genomics **5**(8), 648–654 (2004)
16. Lindberg, C.: The Unified Medical Language System (UMLS) of the National Library of Medicine. J. Am. Med. Rec. Assoc. **61**(5), 40–42 (1990)
17. Fung, K.W., McDonald, C., Srinivasan, S.: The UMLS-CORE project: a study of the problem list terminologies used in large healthcare institutions. J. Am. Med. Inform. Assoc. JAMIA **17**(6), 675–680 (2010)
18. Bodenreider, O., Nguyen, D., Chiang, P., et al.: The NLM value set authority center. Stud. Health Technol. Inform. **192**, 1224 (2013)
19. Quinlan, J.R.: Induction of decision trees. Mach. Learn. **1**(1), 81–106 (1986)
20. Fürnkranz, J.: Separate-and-conquer rule learning. Artif. Intell. Rev. **13**(1), 3–54 (1999)
21. Hearst, M.A., Dumais, S.T., Osman, E., Platt, J., Scholkopf, B.: Support vector machines. IEEE Intell. Syst. Their Appl. **13**(4), 18–28 (1998)

22. Lemeshow, S., Sturdivant, R.X., Hosmer, D.W.: Applied Logistic Regression. Applied Logistic Regression. Wiley, New York (2013)

23. Wang, Y.X., Zhang, Y.J.: Nonnegative matrix factorization: a comprehensive review. IEEE Trans. Knowl. Data Eng. **25**(6), 1336–1353 (2013)

24. Schmidhuber, J.: Deep learning in neural networks: an overview. Neural Netw. **61**, 85–117 (2015)

25. Wang, W.Y., Mazaitis, K., Cohen, W.W.: Structure learning via parameter learning. In: Proceedings of the 23rd ACM International Conference on Information and Knowledge Management, pp. 1199–1208. ACM (2014)

26. Kazakov, D., Kudenko, D.: Machine learning and inductive logic programming for multi-agent systems. In: Luck, M., Mařík, V., Štěpánková, O., Trappl, R. (eds.) ACAI 2001. LNCS (LNAI), vol. 2086, pp. 246–270. Springer, Heidelberg (2001). https://doi.org/10.1007/3-540-47745-4_11

27. Kaelbling, L.P., Littman, M., Moore, A.: Reinforcement learning: a survey. J. Artif. Intell. Res. **4**, 237–285 (1996)

28. Džeroski, S., De Raedt, L., Driessens, K.: Relational reinforcement learning. Mach. Learn. **43**, 7–52 (2001). Kluwer Academic Publishers, The Netherlands

29. Tadepalli, P., Givan, R., Driessens, K.: Relational reinforcement learning: an overview. In: Proceedings of the ICML-2004 Workshop on Relational Reinforcement Learning, pp. 1–9 (2004)

30. SEER-MHOS. http://healthcaredelivery.cancer.gov/seer-mhos/

31. Goldberger, A.L., Amaral, L.A.N., Glass, L., Hausdorff, J.M., Ivanov, P.Ch., Mark, R.G., Mietus, J.E., Moody, G.B., Peng, C.-K., Stanley, H.E.: PhysioBank, PhysioToolkit, and PhysioNet: components of a new research resource for complex physiologic signals. Circulation **101**(23), e215–e220 (2000). http://circ.ahajournals.org/content/101/23/e215.full

32. Johnson, A.E.W., Pollard, T.J., Shen, L., Lehman, L., Feng, M., Ghassemi, M., Moody, B., Szolovits, P., Celi, L.A., Mark, R.G.: MIMIC-III, a freely accessible critical care database. Sci. Data **3** (2016). https://doi.org/10.1038/sdata.2016.35

33. Michalski, R.S., Larson, J.: AQVAL/1 (AQ7) User's Guide and Program Description, Report No. 731, Department of Computer Science, University of Illinois, Urbana, June 1975

34. Wojtusiak, J.: Recent advances in AQ21 rule learning system for healthcare data. In: American Medical Informatics Annual Symposium, Chicago, November 2012

35. Wnek, J., Michalski, R.S.: Hypothesis-driven constructive induction in AQ17-HCI: a method and experiments. Mach. Learn. **14**(2), 139–168 (1994)

36. Bloedorn, E., Michalski, R.S.: Data-driven constructive induction. In: IEEE Intelligent Systems, Special issue on Feature Transformation and Subset Selection, pp. 30–37, March/April 1998

37. Michalski, R.S.: ATTRIBUTIONAL CALCULUS: A Logic and Representation Language for Natural Induction, Reports of the Machine Learning and Inference Laboratory, MLI 04-2, George Mason University, Fairfax, April 2004

38. Wojtusiak, J.: Semantic data types in machine learning from healthcare data. In: International Conference on Machine Learning and Applications (ICMLA), Florida, December 2012

39. Wojtusiak, J., Michalski, R.S., Simanivanh, T., Baranova, A.V.: Towards application of rule learning to the meta-analysis of clinical data: an example of the metabolic syndrome. Int. J. Med. Inform. **78**(12), e104–e111 (2009)

40. Michalski, R.S., Wojtusiak, J.: Reasoning with missing, not-applicable and irrelevant meta-values in concept learning and pattern discovery. J. Intell. Inf. Syst. **39**(1), 141–166 (2012)

41. Elixhauser, A., Steiner, C., Harris, D.R., Coffey, R.M.: Comorbidity measures for use with administrative data. Med. Care **36**(1), 8–27 (1998)
42. Kaufman, K., Michalski, R.S.: A method for reasoning with structured and continuous attributes in the INLEN-2 multistrategy knowledge discovery system. In: Proceedings of the Second International Conference on Knowledge Discovery and Data Mining (KDD-96), Portland, OR, August 1996, pp. 232–237 (1996)
43. Amemiya, T., et al.: Activities of daily living and quality of life of elderly patients after elective surgery for gastric and colorectal cancers. Ann. Surg. **246**(2), 222–228 (2007)
44. Agborsangaya, C.B., et al.: Health-related quality of life and healthcare utilization in multimorbidity: results of crosssectional survey. Qual. Life Res. **22**(4), 791–799 (2013)
45. Taneja, S.S.: Re: impact of age and comorbidities on longterm survival of patients with high-risk prostate cancer treated with radical prostatectomy: a multi-institutional competing-risks analysis. J. Urol. **189**(3), 901 (2013)
46. Vissers, P.A., et al.: The impact of comorbidity on Health Related Quality of Life among cancer survivors: analyses of data from the PROFILES registry. J. Cancer Surviv. **7**(4), 602–613 (2013)
47. https://www.hcup-us.ahrq.gov/toolssoftware/ccs/ccs.jsp

Using Ontologies to Access Complex Data: Applications on Bio-Imaging

Cong Cuong Pham[1], Nada Matta[2(✉)], Alexandre Durupt[1], Benoit Eynard[1],
Marianne Allanic[4], Guillaume Ducellier[2], Marc Joliot[3], and Philippe Boutinaud[4]

[1] Sorbonne University, University of Technology of Compiegne,
Department of Mechanical Systems Engineering, UMR CNRS, 7337 Roberval,
CS 60319, 60203 Compiegne Cedex, France
`{phamcong,alexandre.durupt,benoit.eynard}@utc.fr`
[2] University of Technology of Troyes, 12 Rue Marie Curie, 10010 Troyes, France
`{nada.matta,guillaume.ducellier}@utt.fr`
[3] GIN UMR 5296, CNRS, CEA, Bordeaux University, Case 71,
146 rue Léo-Saignat, 33076 Bordeaux Cedex, France
`marc.joliot@u-bordeaux2.fr`
[4] Cadesis, 37 rue Adam Ledoux, 92400 Courbevoie, France
`{mallanic,pboutinaud}@cadesis.com`

Abstract. Information Systems, used to share information, lead to the growth of heterogeneous data and then the dependencies between them. Thus, the links and dependencies among heterogeneous and distributed data are more and more complex during daily activities of users (researchers, engineers, etc.). Our contribution is to propose a methodology to facilitate the exploitation (interrogation and sharing) of complex data in an organization. The system, we propose, tends to mix semantic approach with data management.

Keywords: Knowledge sharing · Data management · Bio-Imaging · PLM

1 Introduction

Information Systems, used to share information, lead to the growth of heterogeneous data and then the dependencies between them. Thus, the links and dependencies among heterogeneous and distributed data are more and more complex during daily activities of users (researchers, engineers, etc.). The data exploitation (interrogation and sharing) has to be adapted to the context of large data and complex dependencies. To overcome this current inconvenience, more and more research works have investigated and studied the Semantic Web (SW) concepts and techniques to give a cognitive access to information and data. Our contribution is to propose a methodology to facilitate the exploitation (interrogation and sharing) of data in an organization. Techniques to handle heterogeneous and complex data structuring are defined in order to support the dynamic evolution of this type of data. A request system using ontology and semantic mechanism

E. Mercier-Laurent and D. Boulanger (Eds.): AI4KM 2016, IFIP AICT 518, pp. 19–35, 2018.
https://doi.org/10.1007/978-3-319-92928-6_2

is also developed in order to offer a user friendly data management system. This work is defined for Bio-Imaging domain.

2 Related Work

Data access or querying data is the data search process of users to answer a specific question. It is an important function of any information system. From very first development of database technologies, querying data has been a function often dedicated to Information and Technology (IT) experts. Providing non-IT people like end-users with an efficient way to query database, semantic query for instance, was always a challenged topic. In current BMI data management system, some methodologies have been proposed to enable the data access of end-users.

Riazanov et al. [17] proposed a semantic querying of relational data for clinical intelligence. For the authors, *"self-service ad-hoc querying of clinical is problematic as it requires specialized technical skills and the knowledge of particular data schemas"*. A semantic querying allows end-users (clinical researchers, surveillance practitioners, health care managers…) to formulate queries in terms of domain ontologies which are more understandable than data schemas. User queries are then transformed to the one on the data sources by using a mapping between ontologies' terms and data sources.

LORIS [3] is a web-based data management system for multi-center studies from data acquisition to processing and dissemination. The main querying function of LORIS is Data Querying Gui (DQG). By using a web-based interface, DQG *"allows researchers to design, execute, and save queries in a simple and intuitive manner, without having to write complex SQL queries"*. Ping et al. [15] also developed a web-based data-querying tool using ontology-driven methodology and flowchart-based model. The former was used to formulate the query task through a Protégé plugin, the later executes queries and presents the result through a visual and graphical interface.

In the BIRN [11], BIRN Mediator helps users make easily a query to a collection of data sources (such as relational or XML databases, or web services) by provide them with a single consistent virtual schema while the actual data resides in remote sources under their original schemas. *"The mediator maps the sources schemas to the common domain schema, using declarative logical formulas, transparently to the user. The mediator provides the most recent data available, since the user queries are translated into source queries and executed at the sources in real-time"*.

All these proposed methodologies have been based on the well-known approach called Ontology-Based Data Access (OBDA). This approach uses ontology, the pivot component of the Semantic Web (https://www.w3.org/standards/semanticweb/) to enable the semantic data access of end-users by provide them with a semantic representation of data. In this chapter, we propose to use this techniques in order to support knowledge sharing in Bio-Imaging at Gin Lab.

3   Knowledge Sharing in Bio-Imaging

Several researchers in Bio-Imaging need to share their results and data they use. They manipulate heterogeneous data: human information, brain images, diseases descriptions files. They produce more images, statistics data, diseases and brain descriptions. Information grown very quickly and they need a support that handle data evolving than classic database cannot tackle.

Bio-Imaging researchers in GIN lab has been interviewed in order to identify their real needs and their difficulties to share and use information using existing information system during daily activities. Most of scientists have difficulties in data querying and they almost cannot accomplish this task without helps of database technicians. In fact, their data base in very complex; there is a lot of links between data and it is not very easy for them to discover the data base structure in order to understand where a data can be inserted. Otherwise, the data names in the database do not reflect their use. So, when they want to obtain information from data base, they describe their needs, and a technician try to answer them. The problem is on the interaction with database. They need to use incremental requests in order to discover existing data and apply some process on it. Sometimes, they repeat process techniques already done by other researcher; they need to discover their colleague results and use them. Finally, technician cannot update the Database because of its complexity.

Currently, they extract a copy from data base in Excel files (example Fig. 1) and each researcher deal with his excel file. The problem is each one, use specific name of the same data. Results are not shared and used. There is a duplication of the database and its structure. As conclusion, there is no information sharing between researchers due of the complexity and the dynamic updating of database.

Atyp PROD 3 classes de Joinall corrected 18juin14.jmp	sexe	PMa2	SPA	MEM	LEX	Résidu SPA	Résidu MEM	Résidu LEX
Strong-Atypical	H	G	1,439196	0,5526534	0,90757414	0,916525	0,4982332	0,83276172
Strong-Atypical	F	G	-0,14837	1,0057614	-0,08440699	-0,065787	0,6987699	-0,23853232
Strong-Atypical	H	G	1,728121	0,6725831	0,660170273	1,154245	0,3789546	0,26133259
Strong-Atypical	F	G	-0,73354	0,0696222	0,454182742	-0,499536	-0,0082888	0,5418664
Strong-Atypical	F	G	-0,19622	0,6950712	0,534552296	0,101046	0,7243083	0,74040146
Strong-Atypical	F	G	-1,33302	0,5477783	0,473134249	-1,091614	0,4742563	0,56244418
Strong-Atypical	F	G	0,078464	-0,033308	-0,01828836	0,208926	-0,3118861	-0,16188571
Strong-Atypical	H	G	1,220558	-0,532794	0,903145343	0,842473	-0,5710077	0,7558743

Fig. 1. Example of data excel file

To answer these problems, we propose:

- To organize data in incremental way, basing on the main object. In the Bio-imaging case, it is the studied person (named as subject). So data identified and produced concerns this subject.

- To use a database management system that enable easily to manage and update this type of data organization. For instance, in our application, we use Product Life Cycle Management Tool (PLM) in order to handle this type of organization. In fact, structure of data in PLM is organized as the evolution of the product from the idea to the concept and product characteristics. Adding of that, several forms of data can be supported: data, images, text files, etc. We know that different data management systems exists (SQL, NoSQL, etc.) and data organizations (object, relational, etc.) [16]. The PLM system, we used is based on relational data base. We prefer to use a PLM system in which the organization of data correspond to the need of Gin lab bio-imaging researchers; keeping data as an evolution of studies around a subject.
- To develop a request interface as a support of domain vocabulary and links to database. So, a domain ontology has been defined and a request system has been developed in order to handle knowledge sharing between bio-imaging researchers at Gin-lab.

Figure 2 illustrates the needs of knowledge sharing at Gin-Lab

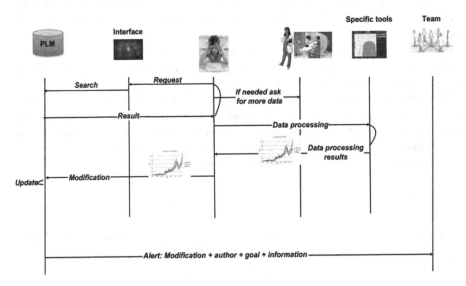

Fig. 2. Needs of Bio-Imaging researchers

4   Structuring Data

As mentioned above, complex data will be structured as relations of a main object. In Bio-imaging studies, the subject (person who follows a medical protocol) is central. So data are structured as to subject as follows:

- subject: number, name, birthdate
- exam: investigator, type, tool, protocole, results
- process: investigator, type, tool, results, references

To support the dynamic evolution of data, PLM TeamCenter tool are used.

4.1 Product Lifecycle Management

The Product Lifecycle Management (PLM) systems integrate constantly all the information produced throughout all phases of a product's lifecycle to everyone in an organization at every level (managerial, technical…) [19]. This type of tools are developed to help product designers and to provide a traceability support of the evolution of n artefact from the requirements to the product use and even recycling (Fig. 3).

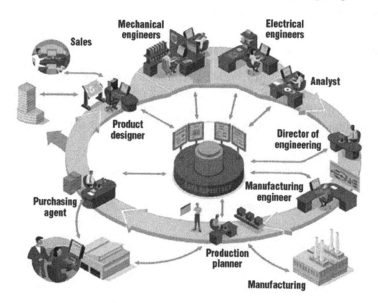

Fig. 3. PLM functions.

We can figure some key advantages of PLM systems [10]:

- Establishing an effective PLM system reduces the enormous data resources to a coherent data flow, avoids redundancies and heterogeneities.
- PLM enables the collaboration through distributed and virtual/extended enterprises (workflow and process management, communication and notifications, secure data exchange…)
- PLM permits the product structure and its evolution management during different steps and track-performed modifications tracking.
- PLM is a mature solution to tackle the heterogeneity, growth and complexity of the data and its processing methods as well as some of the traceability and confidentiality issues.

So, PLM system brings together: Products, service, activities, processes, people, skills, data, knowledge, procedures and standards. It provides an efficient solution to handle the complex and heterogeneous data resources and a mature method to track the evolution and modification of these data.

However, along with these advantages, it also exists some issues:

- Lack of strong stakeholders, ICT tools as well as a common standard between PLM systems causes data integrity problems and limits the access to and sharing of product information and knowledge distributed.
- Another issue of PLM community is the increasing of need for product lifecycle knowledge capitalization and reuse in order to reduce time and cost.
- Database exploitation requires a good understanding of database structure as well as data model especially in the context where the data is heterogeneous and the links and dependencies among data are complex.

4.2 Data Representation in TeamCenter

Figure 4 presents the BMI-LM (Bio Medical Imaging Life cycle Management) data model used in the PLM "TeamCenter 9.1" [1]. By adopting PLM solutions in the context of Bio-Imaging, this PLM-oriented data model covered the whole stages of a BMI study from specifications to publications and enabled the flexibility in data management.

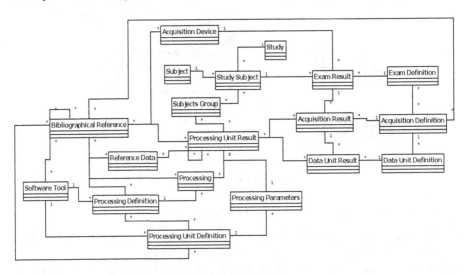

Fig. 4. BMI-LM data model implemented in TeamCenter 9.1

BMI-LM contains three types of objects: *Result* objects (Exam, Acquisition, Data Unit, Processing), *Definition* objects (Exam, Acquisition, Data Unit, Processing) and Reference objects (Bibliographical, Data). "*Definition*" concepts have been used in order to enable the reuse of data. For example, all the *Processing results* computed by using the same *Acquisition device* and the *Processing parameter* can be attached to the same corresponding *Processing definition*.

The classification (Fig. 5) has been built based on the data model. From that, BMI data have been classified into branch, classes and subclasses. The classification allows a specific class to be added to a generic item (object in the data model). In comparing with the data model, the classification and its attributes are easier to modify for user than objects attributes, it is good to fit the model flexibility requirement and also for the appropriation of the database by user [1].

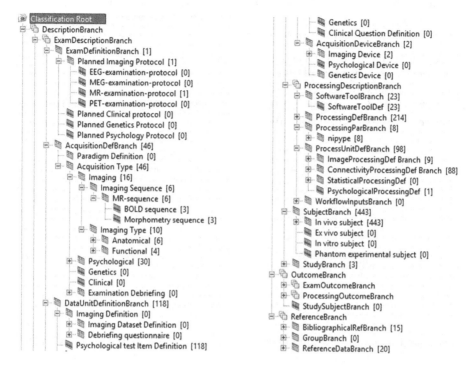

Fig. 5. Classification in corresponding with the BMI-LM data model

In this classification, the nature of data is repeated. In fact, we have images classes as results and as entry data. It is the same for processes, descriptions, etc. The low-level expression of UML schema and the complex relations among classes in the classification also brings difficulties for users in querying the database. To overcome this issue, we build an ontology, which bases on both of data model and classification. This ontology shows logic representation of information in Bio-Imaging and it provides an overview of concepts in the data model and the relationship among them but now represented in a natural language, and therefore it allows end-users to create a query close to his reasoning.

5 Data Access Using Ontology

The concept «ontology» has been used a long time ago in different communities. In the area of computer sciences, "an ontology is a special kind of information object or computational artifact" [9]. In 1993, Gruber [8] defined an ontology as an "explicit specification of a conceptualization" while four years later Borst [2] defined it as a "formal specification of a shared conceptualization". This definition implied that the conceptualization should be readable by machine (formal format) and should be expressed a shared view between several parties, that means a consensus rather than an individual view. Merging Gruber's and Borst's definitions, Studer [21] stated that: *"An ontology as a formal explicit specification of a shared conceptualization"* (Fig. 6). The use of ontology brings some benefits:

- Support communication and cooperation among systems: Ontology enables interoperability, and integration of heterogeneous data sources.
- Enable the knowledge sharing and knowledge reuse.
- Enable content-based access and provide automated services based on machine (the key component of the Semantic Web).

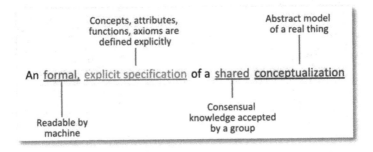

Fig. 6. Definition of ontology [21].

Data access using ontology or Ontology-based data access (OBDA) is a new paradigm for data integration and accessing data sources with complex structure [13] It aims to provide end-users with a semantic access to databases by using a three-level architecture [14] containing:

- Conceptual layer (ontology layer),
- The data sources,
- The mapping between the ontology and the data source.

The ontology acts as a mediator between user and the data source. Its aims to provide user with a semantic representation of data sources by using a set of concepts in the domain of interest and all relations among them. The data sources are the repositories of data stocked in a relational or non-relational databases. The mapping layer maps the domain concepts to the data sources.

The queries formulated by users using concepts of ontology are translated to the one on the data sources by using this explicit mapping (Fig. 7).

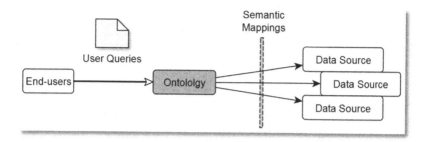

Fig. 7. Architecture of OBDA systems

This principle are used to define an Ontology Based System in order to support data access in Bio-Imaging at Gin lab. Before describing the architecture of this system, let-us define the used ontology.

5.1 Bio-Imaging Ontology Definition

There is a lot of ontology defined in medicine, for instance, in oncology, neurology [6]. But, little work study bio-imaging representation. Gibaud et al. [7] define concepts used in Bio-imaging like: Dataset, Processing, Investigators, Medical Image files, Equipment, subjects, etc. (Fig. 8).

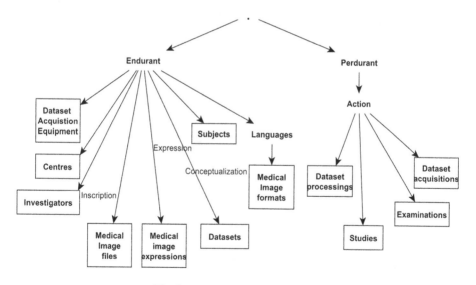

Fig. 8. Ontology for Bio-Imaging.

We adapt this ontology corresponding of the logic use of information in database. Information belongs to three major categories: **Tools**, **Data**, and **Process** (Fig. 9).

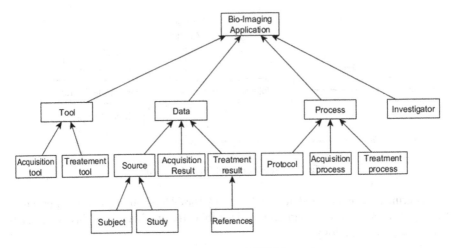

Fig. 9. Conceptual tree of GIN lab ontology

Several relations exist between these concepts like **Use, Follow** and **Provide** (Fig. 10).

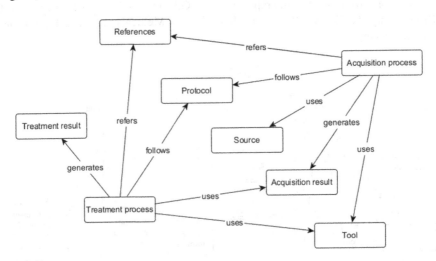

Fig. 10. Conceptual graph of GIN lab ontology

5.2   Ontology - Data Link

Ontologies' low-level concepts have to be linked to data in database, or unless respect the variable name of these data. An inference engine can help to build a data request and generate links using the propagation of relations between concepts (Fig. 11).

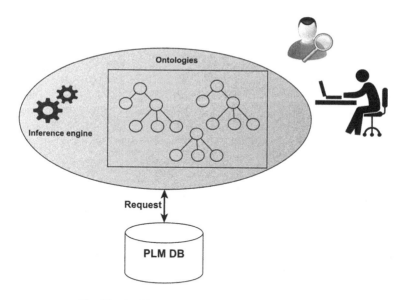

Fig. 11. Architecture of knowledge sharing in PLM

To generate request on database, relations between concepts and data types are defined. For each concept, several data types are linked. For instance, acquisition protocol is related to different protocols in magnetic-resonance-imaging acquisition process. These links are integrated in XML file as interface between Inference engine and database management tool (the TeamCenter in GIN lab).

A request system is then developed in order to provide a user friendly data request system.

Links between Ontology and database model.

5.3 Ontology and Classification Mapping

A table of mapping between ontology concepts and classes is required since only classes of the classification of data are connected to data. Using this table of mapping, the query formulated by users with vocabularies of ontology will be translated to the one that is understandable and executable by a Query Processor.

One concept can be connected with many classes and vice versa. To simple the mapping, firstly we used some visual and interactive tools (Free-mind - http://free-mind.sourceforge.net for instance) to map classes to concepts. The mapping was then transformed into XML and integrated in the query system. Figure 12 illustrates the mapping between concept "image-acquisition-protocol" of imaging ontology (Fig. 12) and all leaf classes in branch "AcquisitionDefinition-Branch/Imaging" of the classification of data.

Fig. 12. Ontology concepts tree and the classification of data mapping.

5.4 Ontology Based Query System

By using ontology tree and ontology graph, ontology-based graph query interface helps users to make a query more easily. Using the ontology tree and ontology graph, users can understand the relationships among concepts and directly choose query parameters. Users also can choose a query in query history to re-execute, modify or complete it. When a user completes his query, our system does as following:

1. Identifying nodes links, following relations in ontologies.
2. Generating an output query in a format understandable and executable by Query Processor, XQuery Engine for example.
3. Executing the output query on PLM data file (.xml or .json format).
4. Results are then visualized as a graph and data in the Interface query.

With the support of these graph, user understand relationships among data in the database, he/she defines the conditions of query according to their purpose. We take here an example of query frequently used by scientists at GIN lab:

"Querying about Brain image results from Retin Treatment using Dash protocol corresponding to men less than 45 years old, left handled and passed exam before January 2013"

If a researcher tries to define a SQL query using PLM related to this question, without using ontology. He/she has to know relations between Study Subject, Exam results, Processing definition and Processing Results data in PLM Database. First, the name of objects in data base are more computer driven called and he has to follow all links in order to identify the ones corresponding to his query which can then be:

Select Image Results from Processing Results Where Processing Protocol = "Dash" and Processing Treatment = "Retin" and from Acquisition Result where Acquisition Date < "January 2013"and from Study Subject where Study Subject Gender = "men", Study Subject Handless = "Left handled" Study Subject Age < "45".

Using Ontology Interface (Fig. 13), she/he can select start by select the concept "Subject", as for concept parameters, selects "gender" and "Age", writes her/his choice (Men and <45 years old). Then she/he selects the concept "Protocol", writes the choice (="Dash"), selects the concept "Treatment Process", choice (="Retin"), and "Treatment results" (File = "Image"), she/he selects concept "Acquisition result" (Acquisition psychology laterality handless = "Left handled") and (Date of Acquisition < "January 2015").

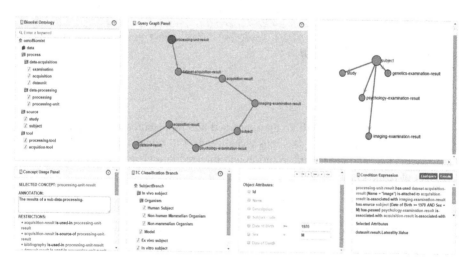

Fig. 13. Ontology-based graph query interface

The inference engine identifies relations between Subject, Acquisition and Process:

- Acquisition "Use" Source
- Protocol "Use" Acquisition
- Treatment "Follow" Protocol
- Treatment "generate" Treatment Result

By subsuming, relations are inherited so, the result of the inference engine will be:
Treatment-Result-Type (=Image) "Follows" Protocol-Name (=Dash) and Protocol-Treatment (=Retin) "Uses" Acquisition-Laterality-Handless-result (=Left handled) "Uses" Subject-Gender (=Men) and "Uses" Subject-Age (<45).

This result query description is then transformed on a TCquery that considers PLM database as a XML file, the output query is transformed to xQuery in order to extract data from that file:

```
declare function local:getQueryObject(){
    let $query := <query query_name='Item...'>
                    <param name='Type' value='ProcessRes'/>
                    </query>
    return Teamcenter:Query($teamcenter, $query)
};
let $processingResults := local:getQueryObject()
for $processingResult in $processingResults
return if(fn:count($processingResult[(@protocol = "Dash") and (@treatment = "Ret-
in")]/F_GIN4_rel_Acquisition/GIN4_Acquisition[@laterality_handless = "Left han-
dled" ]/B_ GIN4_rel_ExamenRes/GIN4_ExamRes[@examDate <= "01 January
2013"]/B_GIN4_rel_ExamRes/GIN4_StudySub [(@ gin4_sex = "M") and
(@gin4_databirth >= "1970")]) >0)
then (processingResult)
else ()
```

The results of the query will be then returned and represented in a graph or a table in the same query interface. Users can visualize data to understand the relationships among them or to find hidden information and inferred knowledge. Figure 14 presents an extract of results represented in a graph.

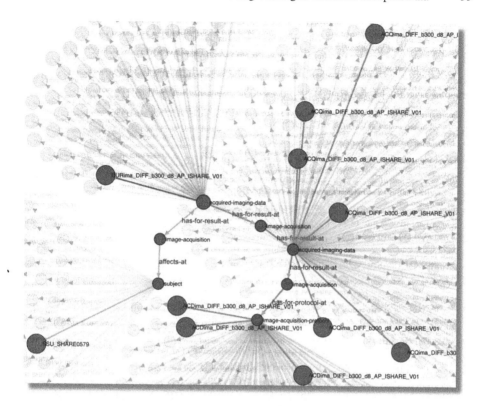

Fig. 14. Graphical representation of results. Click on a node to highlight all related nodes.

6 Conclusion

In this complex and heterogeneous data management problem is studied; How to search data without knowing the data structure, how to discover data defined by actors in an organization, how to share knowledge and data in friendly way. A general approach for ontological model construction and an ontology-base query interface is presented as a solution to tackle the difficulties in querying complex database. Data will be organized around objects. For that, specific data organization systems can be used as PLM. PLM help to organize data as the evolution description of product parts. an architecture and request system are developed in order to build links between data management and semantic system. A use case in Bio-Imaging domain has been also used to illustrate the abilities of our proposed interface.

As future work we will focus on the test of proposed query interface with various queries sets (in Bio-Imaging domain) and engineering design (in PLM). The ontology tree and ontology graph must be also developed to cover all concepts in Bio-Imaging domain. Ontology will be implemented in semantic web language (RDF, SPARQL) in order to use inference engine for information search. We propose to generate an alert system when new data are added with semantic annotation of these data in order to show the reason of their creation.

Acknowledgments. The work presented in the paper is conducted within the ANR (Agence Nationale de la Recherche) founded project BIOMIST.

References

1. Allanic, M., Durupt, A., Joliot, M., Eynard, B., Boutinaud, P.: Application of PLM for biomedical imaging in neuroscience. In: Bernard, A., Rivest, L. (eds.) PLM 2013. IFIP AICT, vol. 409, pp. 520–529. Springer, Heidelberg (2013). https://doi.org/10.1007/978-3-642-41501-2_52
2. Borst, W.N.: Construction of Engineering Ontologies for Knowledge Sharing and Reuse. Universiteit Twente (1997). http://doc.utwente.nl/17864/
3. Das, S., Zijdenbos, A.P., Vins, D., Harlap, J., Evans, A.C.: LORIS: a web-based data management system for multi-center studies. Front. Neuroinformatics **5**, 37 (2012). https://doi.org/10.3389/fninf.2011.00037
4. Davies, J., Duke, A., Technologies, B., Park, A., Ip, I.: OntoSharentoShareaahahake, A., Technologies, B., Pa13
5. Fensel, D., van Harmelen, F., Horrocks, I., McGuinness, D.L., Patel-Schneider, P.F.: OIL: an ontology infrastructure for the semantic web. IEEE Intell. Syst. **16**(2), 38–45 (2001)
6. Gibaud, B., Forestier, G., Benoit-Cattin, H., Cervenansky, F., Clarysse, P., Friboulet, D., Gaignard, A., Hugonnard, P., Lartizien, C., Liebgott, H., Montagnat, J., Tabary, J., Glatard, T.: OntoVIP: an ontology for the annotation of object models used for medical image simulation. J. Biomed. Inf. (JBI), Elsevier (2014). ISSN 1532-0480
7. Gibaud, B., Kassel, G., Dojat, M., Batrancourt, B., Michel, F., Gaignard, A., Montagnat, J.: NeuroLOG: sharing neuroimaging data using an ontology-based federated approach. In: AMIA Annual Symposium Proceedings of American Medical Informatics Association, vol. 2011, p. 472 (2011)
8. Gruber, T.R.: A translation approach to portable ontology specifications. Knowl. Acquis. **5**(2), 199–220 (1993). https://doi.org/10.1006/knac.1993.1008. Special Issue: Current Issues in Knowledge Modeling
9. Guarino, N., Oberle, D., Staab, S.: What Is an Ontology? In: Staab, S., Studer, R. (eds.) Handbook on Ontologies. International Handbooks on Information Systems, pp. 1–17. Springer, Heidelberg (2009). https://doi.org/10.1007/978-3-540-92673-3_0
10. Eynard, B., Gallet, T., Nowak, P., Roucoules, L.: UML based specifications of PDM product structure and workflow. Comput. Ind. **55**(3), 301–316 (2004)
11. Helmer, K.G., Ambite, J.L., Ames, J., Ananthakrishnan, R., Burns, G., Chervenak, A.L., Foster, I., et al.: Enabling collaborative research using the Biomedical Informatics Research Network (BIRN). J. Am. Med. Inform. Assoc. **18**(4), 416–422 (2011). https://doi.org/10.1136/amiajnl-2010-000032
12. Hendriks, P.: Why share knowledge? the influence of ICT on the motivation for knowledge sharing. Knowl. Process Manag. **6**(2), 91–100 (1999)
13. Khamis Abdul-Latif, K., Luo, Z., Song, H.Z.: Modified Query-Roles Based Access Control Model (Q-RBAC) for interactive access of ontology data. J. Inf. Eng. Appl. **4**(7), 82–91 (2014)
14. Lenzerini, M.: Ontology-based data management. In: Proceedings of the 20th ACM International Conference on Information and Knowledge Management, CIKM 2011, pp. 5–6. ACM, New York (2011). https://doi.org/10.1145/2063576.2063582

15. Ping, X.-O., Chung, Y., Tseng, Y.-J., Liang, J.-D., Yang, P.-M., Huang, G.-T., Lai, F.: A web-based data-querying tool based on ontology-driven methodology and flowchart-based model. JMIR Med. Inf. **1**(1), e2 (2013). https://doi.org/10.2196/medinform.2519

16. Ramakrishnan, R., Gehrke, J.: Database Management Systems. Osborne/McGraw-Hill, CA (2000)

17. Riazanov, A., Klein, A., Shaban-Nejad, A., Rose, G.W., Forster, A.J., Buckeridge, D.L., Baker, C.J.: Semantic querying of relational data for clinical intelligence: a semantic web services-based approach. J. Biomed. Semant. **4**, 9 (2013). https://doi.org/10.1186/2041-1480-4-9

18. Sato, H., Otomo, K., Masuo, T.: A knowledge sharing system using XML Linking Language and peer-to-peer technology. IEEE Xplore, 26–27 (2002)

19. Sudarsan, R., Fenves, S.J., Sriram, R.D., Wang, F.: A product information modeling framework for product lifecycle management. Comput. Aided Des. **37**(13), 1399–1411 (2005)

20. Small, C.T., Sage, A.P.: Knowledge management and knowledge sharing: a review. Inf. Knowl. Syst. Manag. **5**, 153–169 (2006)

21. Studer, R., Richard Benjamins, V., Fensel, D.: Knowledge engineering: principles and methods. Data Knowl. Eng. **25**(1–2), 161–197 (1998). https://doi.org/10.1016/S0169-023X(97)00056-6

22. Yoo, D., No, S.: Ontology-based economics knowledge sharing system. Expert Syst. Appl. **41**(4), 1331–1341 (2014)

23. Zhang, J., Liu, Y., Xiao, Y.: Internet knowledge-sharing system based on object-oriented. IEEE Xplore **1**, 239–243 (2008)

Dynamic Ontology Supporting Local Government

Mieczysław Owoc[✉], Krzysztof Hauke, and Katarzyna Marciniak

Wroclaw University of Economics, Wrocław, Poland
{mieczyslaw.owoc,krzysztof.hauke,katarzyna.marciniak}@ue.wroc.pl

Abstract. Nowadays in information society we face a huge amount of data and information sharing on the Web by customers of public administrator sectors. Services performed for customers from the nature are very flexible – so preparation ontology for such goals must include dynamic aspects. The goal of this paper is creation dynamic ontology for relationships type of C2C (Citizen to City). It is especially important in knowledge management at the local level of administration in case of monitoring uncompleted matters, evaluation of effectiveness of clerks representing selected departments or in decisions of administration restructuring. In this paper we propose a dynamic-oriented knowledge model, designed to distribute and manage urban knowledge as a representative of local self-government. Ontology including mentioned knowledge management supporting is prepared using sophisticated Ontorion Fluent Editor.

1   Introduction

Description of the processes that citizens participate in modern society is very complex and cover all areas of human activities. On the other hand relationships between ordinary people (in fact customers) and public administration at different levels of ruling are essential for functioning of any society. Good ontology allows for better understanding of the mentioned processes and creates opportunity to manage efficiently social life in different areas.

There is relatively small number of research investigating nature of relationships between citizens and governments at the state as well as at the local levels (citizen-to-city: C2C). Despite of certain similarities (mostly participation of two essential components: people and government body) there are several significant features important for the relationships C2C: common area of living and acting, communication, education and many others. Basically, research in this area is focused on an idea of smart city where more ambitious solutions are offered (see: Batty 2013; Townsend 2014). As a result so-called urban knowledge can be introduced as general platform adequate for city society.

The main paper goal is presentation of ontology describing main components of the defined domain including dynamic aspects of its functioning. Therefore the following sections are present in this paper: discussion about the urban knowledge, demonstration of main areas covering C2C model, presentation of initial version of ontology with active rules essential in the model and conclusion with synthesis of research findings.

2 Specialty of Urban Knowledge as the Informational Infrastructure Essential in C2C Model

Climate and demographic changes, limited resources, growth of population, urbanisation, increasing importance of information and development of information technologies are forcing large and medium-sized cities to make changes in every area of their operational functioning. Beginning from integrating autonomously functioning ICT platforms through effective energy resources, raw materials and waste managing and ending with developing dialogue with citizens and making physical infrastructure changes.

The main aim of such upgrade is not only to achieve positive economic impact on a city, region, or a country, but mainly to prepare for meeting future needs of civilisation. Nowadays they are generated by the society which in fact is classified as an information society. Solutions used in cities all over the world (Amsterdam, Copenhagen, Montpellier, New York, Singapore, Wrocław, etc.) are no longer sufficient to provide proper communication between its users - citizens, but has become an integral part of the present civilisation infrastructure. Such situation forces city governances into using effective management in all different branches of their urban economies. That will lead to the development of higher levels of efficiency, interactivity, flexibility, accountability essential to adapt to the rapid pace of changes which, indeed, is the ideological basis of well-known and used in the world concept of smart city.

Main factor of describing city as a kind of smart is intelligent management and knowledge management. It means that if the decision-making bodies make decisions regarding the development of the city they must take into consideration contractual six segments, aspects of the agglomeration: smart economy, people, governance, mobility, environment and living (Chourabi et al. 2012). Figure 1 illustrates the pattern of relations occurring between so called sectors of the modern city. This classification describes how strong intersectoral cooperation should be provided to ensure effective decision making processes based on relevant knowledge. Such visualization shows also how crucial can be dependence on Information and Communication Technologies. The purpose of the existence and operation of ICT infrastructure is therefore necessary to integrate key information generated by its users on each field, which is to provide a complete list of requirements, guidelines for maintenance and improvements in aspects: ecological city life of its citizens, public safety, public services and the operation of any commercial and industrial activities.

To ensure the effectiveness and efficiency of smart city it is crucial to apply all dedicated solutions related to appearing problems concerning each and every area of the city taking advantage of implemented ICT solutions. The existence of individual components, as autonomous entities is impossible. That is why it is important for decision-makers of a city to think of a process of making changes as a holistic investment and development. Decisions should be made not only on current information presented by mainly on gathered knowledge and experience.

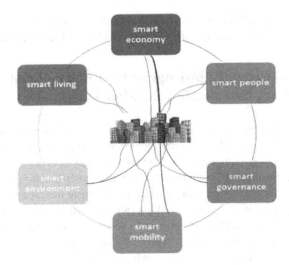

Fig. 1. City as a system.

Looking at the city from the perspective of the whole, complete, living organism allows decision-makers to pay special attention to the integration of urban infrastructures. That mechanism of making changes in one aspect relating results in other aspects make smart city as a "system of the systems".

Wherever processes of decision-making are found, through their algorithmicising, it is possible to use information support, using specially dedicated solutions. Strategic vision for the implementation of information systems in urban areas should be understood primarily through the integration of diverse information systems and customization them for the needs of residents, using all available communication channels and tools.

According to the author's knowledge, a key element for the pursuit of development in line with the reasons stemming from the macro-cities is an urban knowledge. Taking into account the results of enterprises market functioning and the results that achieve through the use of knowledge management processes, the author believes that urban knowledge may be an aspect that will allow raising the competitiveness of cities, seeking to ensure the integration of infrastructure (mainly in the area of information technology, social and governance) and achieving spatial cohesion. Thus, the author believes that the improvement of management processes in the cities, on the basis of knowledge can contribute to the real fulfillment of assumptions Regional Development, the National Spatial Development Concept 2030 in 2030 at the local country.

Knowledge is not only one of the key resources of the enterprise, but also is the foundation, the starting point for determining the city's strategies, particularly for the implementation of management information systems. However, the organization cannot exist without human capital, so to be able to say that the company has knowledge, it is needed to take into account in addition to possession of selected information: skills, experience and qualifications of specialists in selected fields. The combination of these two elements is a complete understanding of each organization. Parallelly, paying attention to the continuity of the process of converting data into information and information into knowledge (see: Albescu et al. 2009).

Because of the persistence of highlighted perspectives knowledge is classified as an information resource and as an element of human capital. Each organization needs both to ensure proper functioning. It is reasonable to say that knowledge is nothing else than "a combination of everything: facts, phenomena and relationships between them, which is consciously perceived and recorded (in any way saved as real entities or conceptual) and can give to others, according to the intention of having knowledge in specific conditions and circumstances to arouse certain behaviours" (see: Bergeron 2003).

Trying to define knowledge as an asset of such complex organization like city it is necessary to investigate more facts. Presented definitions of knowledge in organization are considered as basis but with certain are insufficient. To have holistic description of urban knowledge author consider adding some more features important for defining the intangible (see: Urban Knowledge 2015).

Urban's knowledge, considered as an asset, needs to be of course possible to manage. Based on ongoing processes of city functioning, it was possible to notice, that knowledge of the city is created independently by entities functioning beyond the city's structure. This kind of knowledge must be secured in very specific way, is unique for long time. City's knowledge is created by local authorities, policy makers, social activists, institutions (educational medical, religion, ecological, others), R&D, administration units, business. According to model of smart city (presented in the previous section), authors decide to divide city's knowledge into six aspects – see Table 1.

Table 1. Components of urban knowledge

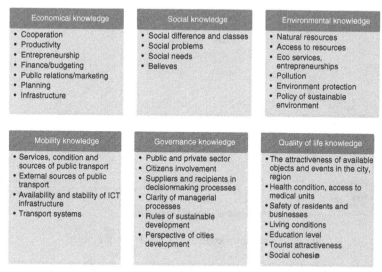

As we can see, urban knowledge can be understood as the summary of: economic knowledge, people knowledge, environment knowledge, mobility knowledge, governance knowledge and living knowledge embracing specified knowledge pieces.

Urban knowledge classified as the asset of the city can be managed. Urban knowledge management can be treated as a specially designed system that helps cities to acquire, analyse the use (re-use) of urban knowledge in order to make faster, smarter and better decisions, so that they can achieve a competitive advantage in case of covering cities stakeholders needs. Urban knowledge management covers management of information, knowledge and expertise available within the city, i.e. mobility, environment, social, managerial, economical by the creation, collection, storage, sharing and use, to ensure the cities future development based on well prepared decision plans. Urban knowledge management also emphasizes the two items related to knowledge management. Both accessible to, the information, experience, staff and their expertise and the technological, where the focus is on codifying knowledge, its acquisition, collection, analysis, storing and sharing at any time, by a specific user. The logical also is the fact that the development of knowledge takes place through the exchange of experiences, analysis, opinion, finding new sources of information, where the information systems are the basis to allow all of the actions.

In addition, each city must be aware that the overall urban knowledge management system cannot be based only on properly chosen technology" – (Tiwana 1999). Because of presented arguments, it is possible to propose Local Governments well-known e-commerce model: Consumer to Consumer (C2C) as adequate for supporting the urban knowledge management as the method of implementing their self-governance. Knowing that C2C is model which de-scribes process of exchanging goods, services between users on third-party business created to facilitate the transactions, authors may assume that C2C in case of urban knowledge may be defined as model of exchanging the information and knowledge (urban) on-line between cities stakeholders by using knowledge sharing platform (Fig. 2).

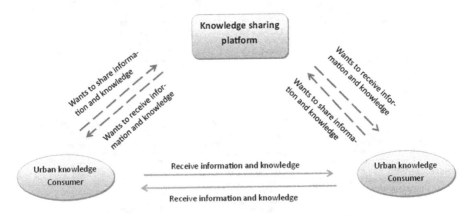

Fig. 2. Urban knowledge in C2C model for the city.

According to presented concept of urban knowledge, its categorization visualized in Table 1 and model of C2C Fig. 2, urban knowledge management or urban knowledge transfer may be supported by information system in form of knowledge sharing platform. User of this platform are in general consumers of urban knowledge (e.g. citizens, local government units, administration officers, representatives of science, entrepreneurs, etc.). Knowledge sharing platform can be the perfect place to support knowledge gathering, storage and sharing for its users.

Urban knowledge sharing platform for can be a compendium of technological information, in accordance with the requirements of the concept of "knowledge grid for urban knowledge management", which means that:

1. Proposed Knowledge Platform will create a place where indicated institutions responsible for the creation of legal, laws and regulations, social workers, institutions interested or involved in the solving the cities problems (universities, research institutions, scientists, the media, journalists), independent entities interested in developing the city, individual specialists will be able to share content about actual city prospering in particular fields.
2. For platform users, platform becomes a compendium of knowledge about the urban knowledge (economy, people, environment, mobility, governance, living).
3. Platform will also provide a place of learning for specific social groups. The platform will include a module for e-learning, where the above group of interests will be able to create educational courses for specific groups of users.
4. Platform will be a place and a tool which will support the knowledge management process in the city. Defining the users of the platform, authorities are already informed about the group of interest in the city. Building the communication with the users ensure knowledge gathering. Using the platform as a tool for support city's management ensures also the integration of ICT infrastructure in the city and make the clarity and order in the city.

Considering development of dedicated knowledge portal at least the following aspects should be taken into account: purpose and audience, technology and tools useful for the creation and maintenance stages.

Platform can enable the integration of the information needs of citizens and all the entities interested. Information needs of citizens and all the entities interested. Table 2 presents potential lists of expectations of implementing C2C model in the city from perspective of chosen urban knowledge consumers: city's authorities, units cooperating inter-sectoral within the city and external stakeholders.

The idea of that platform is to provide the right kind of knowledge related to the particular recipients. In addition, the platform provides users the ability to communicate with each other: both the target group, which will be the citizens of the city, and the target group with the suppliers of knowledge and educational materials. Users will also have the ability to communicate using the tools (i.e., hut, video, mobile, blogs) with the other participants will help to alleviate the barriers between different social groups, and at the same time will help to increase the effectiveness. Platform for manage the urban knowledge (presented in Fig. 2) and in the same time ensure existence of C2C model, can be a compendium of technological information, in accordance with the requirements of the concept of "knowledge grid for urban knowledge management" (Hauke et al. 2015), which means: centralization of any knowledge resources focused on C2C model, inclusion of crucial components (economy, people, environment, mobility, governance, living) and easiness of urban knowledge management.

Table 2. Expectations of chosen urban knowledge consumers in C2C model.

City's authorities	Units cooperating inter-sectoral	External stakeholders
• Indentification of neccessary knowledge and its gaps • Possibilitiy of providing advanced analysis • Shorter time of knowledge acquiring • Improvement of knowledge exchange between users • Possibility of useer experience registering and sharing	• Access to relevant knowledge • Possibility of feeding to knowledge bases • Providing more advanced sector and inter-sector analysis • Easier knowledge and information flow between units • Unifed model of knwoledge gathering and exchange	• Possibilities of providing and receiving more details reports about the city functioning • Access to information about public services • Easier access to marketing activities provided by the cities

Considering development of dedicated knowledge portal at least the following aspects should be taken into account: purpose and audience, technology and tools useful for the creation and maintenance stages.

3   Areas of Activities of a Local Self-Government

Local Self-Government (LSG) can be identified with local government as a specific form of public administration oriented on some commune. There are several features of this sort of administration; widely accepted are: autonomy, sovereignty, home ruling in every case acting in a small scale and focusing on local matters.

According to the aims of a Local Self-Government all domains connected to local community should be included to its activities. Therefore the regulations in the following areas must be present:

- society life (e.g. people registering, people mobility, education at the lower levels, paying taxis and others identified with personal needs of habitants),
- infrastructure maintaining (in terms of services of crucial means: water delivering, sewers services, electricity and gas supplying, local communication and the like),
- business services (business evidence, office renting, taxation, commercialization etc.).

Analysing different city portals we discover variety solutions addressed to local society and other potential customers (business and travellers for example). Our research is focused on representatives of three categories of cities: megacity (with more than 10 million habitants), middle city (range 1-10 million habitants) and relatively small city (less than 1 million habitants). Additionally we selected cities from different parts of Europe, thus three cities were chosen: London, Berlin and Wroclaw. As a result considering local administration we discovered the following divisions representing administration offices of different cities in presented in Table 3.

Table 3. Examples of divisions in selected European Cities

Urban knowledge	Activity Areas	London	Berlin	Wroclaw
Social, Quality of life	Service of habitants	Communities, Migrants and Refugees, Older people	Labor Market, Personal Data Evidence	Registering and archives of Personal Data
Social	Education	Education and youth	Living Studying and Working	Enrollment, scholarships
Social, Quality of life	Health, Culture	Healthy inequalities, Healthy environment, Sport, Regeneration, Arts and Culture	Services and Contacts	Health information, Services, Cultural Institutions Mgmt
Mobility, Social	Communication	Transport	Public transport	Car and Driving Licence, Public transport
Government, Environment	Infrastructure	Housing and land, Renting	Commercial Trade and Real Estate	Real Estate Mgmt, Municipal Greenery
Social, Quality of Life	Safety	Policing and Crime	Services and Contacts	Municipal police
Economical	Investment, Business Services	Business and Economy, Research and Analysis	Berlin as Economic Centre, Investing in Berlin	Urban Investment, Public procurement

Presentation of Local Self-Government organizational aspects we may conclude that all urban knowledge and activity areas are present in particular cities. Some of divisions are specific for megacity (London Immigrants and Refugees Departments), some are strictly oriented on city affairs (Berlin as Economic Centre) while specific nature of infrastructure for smaller city (Wroclaw Municipal Greenery). In all cases C2C model expresses complex nature of entity-government relationships including its dynamical aspect.

4   Ontology for Local Self-Government Knowledge Management

Before the municipality as a basic unit of local self-government are facing new challenges. It plays an integral role in relation to the inhabitants and entities operating in the area. The implementation of the mission of the municipality is carried out on the basis of conditions resulting from state policy in relation to the given areas, local conditions, which are the result of activities of municipalities themselves. In this process, you should not forget about the stakeholders who live in the municipality or do business. These are the entities that will represent the new quality we are dealing with, which is called attractiveness. In order for the municipality to be regarded as attractive, residents and businesses must actively participate in it. Their effectiveness is supported, among other things, by information technology.

Local government at the commune level has experienced a period of very dynamic changes over the last few years. These changes are influenced by many factors. Among them, information technology plays an important role. The opportunities that web services now offer should be adapted to the management of the municipality. The changes that result from the use of information technology are adapted by business organizations. There may be some delays in the implementation of modern solutions that support management processes. This includes the nature of the impact and the way local government units are managed as a whole. The inflexible mechanisms of information flow and the monopoly of decision-making for the stakeholders of the municipality (inhabitants, companies, public benefit organizations) have led to the status of municipalities (see Fig. 3). Stakeholders do not always accept decisions made by municipalities. This critical opinion arises mainly from the ignorance of activities in a given area and the lack of contact with those who take on behalf of the municipality.

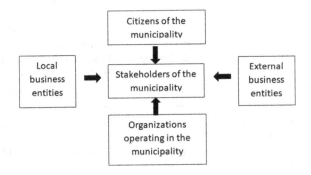

Fig. 3. Stakeholders of the municipality

The mission of the public administration is resident supporting in the defined area. Regardless of geographic location, public administration has to solve the problems of its inhabitants. Public administration is a complex organ, which can distinguish the following categories:

- inhabitants,
- cases referred to the office under defined conditions,
- competent authority of the public administration, which solves particular problems of the defined matters.

Public administration departments in order to complete inhabitant's matters use: domain knowledge and in particular cases defined earlier procedures. Knowledge and procedures that result from established law and current regulations in force in the territory.

From the organizational point of view public administration it is divided into divisions entitled to complete entities (inhabitants, Business, other institutions) matters e.g.: impact areas of public administration, offices, and departments.

The Fig. 4 shows infrastructure of Local Self-Government activities.

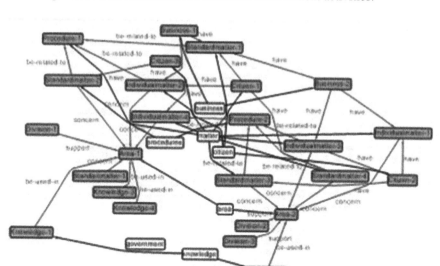

Fig. 4. Main categories of Local Self-Government activities

The particular categories are the components of the city public administration infra-structure. Relationships among particular components can be formulated as direct on indirect. Faced with the direct relationships between its elements are operators, as steps. Below we will discuss examples of ontologies occurring between the elements of the environment of public administration (Fig. 5 and 6).

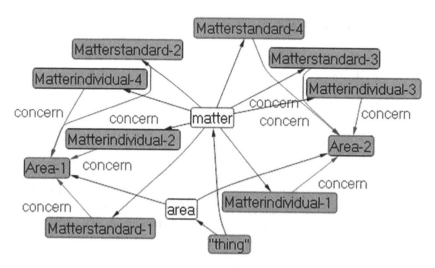

Fig. 5. Ontology - matter "concern" area

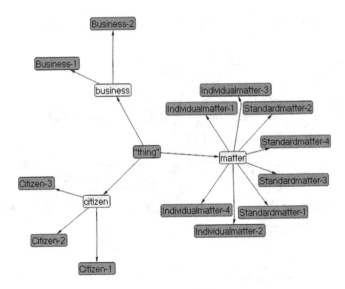

Fig. 6. Ontology - matter "concern" area

Example 1
Operator: concern; elements: matters (standard and individual matters), areas

Example 2
Operator: has; elements: matters (standard and individual)), citizens, business
  Example of question to ontology:

Q1. Who-Or-What uses Area-2?
Ans. Knowledge-2
Q2. Who-Or-What has StandardMatter-4?
Ans. Citizen-1
Q3. Who-Or-What is-related-to Procedure-1?
Ans. StandardMatter-1
    StandardMatter-2
    IndividualMatter-1
    IndividualMatter-2
Q4. Who-Or-What is-related-to Procedure-1 and concerns Area-2?
Ans. Matterindividual-1

The presented ontology allows to organize the components of the municipality. The consequence of this may be the implementation of an active rule. Thanks to this traditional management in the municipality can be implemented automatically. Many tasks in the municipality are repetitive. If we can see this repetition as an algorithm, then it means that we can apply the rules without any problems. Active rules are determined by four operations (Fig. 7):

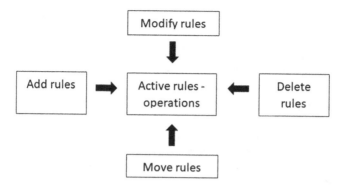

Fig. 7. Active rules – operations.

- add,
- modification,
- moving,
- deleted.

However, for two operations be very careful. This is an operation to move and delete an object. In this case, there can be no such approach, which does not take into account the mechanism of recording the history of the opiate (Fig. 8).

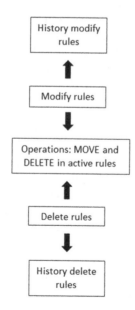

Fig. 8. Operations MOVE and DELETE in active rules.

Active rules:

1. Example KnowledgeInsert
   - If citizen is new then add to citizen.
   - If business is new then add to business.
   - If new law is created then add to use.
   - If new local politics then add to use.
   - If business is new then add to taxes.
   - If knowledge is new then knowledge add to Area-1 or Area-2.
   - If matter is incomplete then matter is back to citizen.
   - If matter is Division-1 then use Procedure-1 or Pro-cedure-2.
2. Example KnowledgeDelete
   - If matter has status = TRUE then matter move to archives.
   - If knowledge is outdated then knowledge delete from actual knowledge.
   - If citizen it has taken care of all matter then citizen delete from ontology.
   - If procedure unresolved then procedure delete from ontology.
   - If business is liquidated then delete from taxes.
   - If the laws are outdated then laws move to archives.
   - If business gone bankrupt then move to archives.
   - If business gone bankrupt then business from ontology.

To develop ontologies have been selected basic objects handled by the public administration. For example, a big number of objects selected to develop ontologies denotes the complexity in which the public administration has to carry out its tasks. It should be stressed - as a result of the mission of public administration - that all tasks performed by the public administration must be coherent with all regulations formulated for the society (see: Breuker et al. 2009). In the case of addressing particular matters for the LSG there are different procedures prepared for variety customers (citizens or representing business).

Because of their mission, they do not compete with each other so much. However, mechanisms are created that should motivate the activities of the municipality for the needs of the environment. The community's stakeholder setting will make the community attractive. In this case you can talk about the development spiral of the municipality. Increasing the attractiveness of the municipality results in a greater interest in the activities of the municipality by residents and businesses, which is on its territory. Through the economic mechanisms, that is, taxes on business activities by the business entity and opinion of the inhabitants of the municipality and also their taxes increase the per capita position of the municipality. This in turn translates into the interest of the municipality concerned by stakeholders who want to live in its territory or do business. At present, such activities make it easier for municipalities to raise funds from central funds or from the European Union.

5   Conclusions

Analysing services essential in C2G models we discover specific aspects and context when customers become citizens and government is limited to local area. The main findings of the research are:

- C2C model can be defined as sub-model of C2G model;
- Components of presented ontology are partially similar despite of different areas of activities (comparison C2C abd C2G);
- Dynamic aspects of ontology must be represented in more advanced ways as active rules;
- Actually the tool used in the research offered very limited way of expression dynamic knowledge.

In the nearest future we'll try to analyse and evaluate other tools for the defined C2C model.

References

Albescu, F., Pugna, I., Paraschiv, D.: Cross-cultural knowledge management. Inform. Econ. **13**(4), 39–50 (2009)

Batty, M.: The New Science of Cities. MIT Press, Cambridge (2013)

Bergeron, B.: Essentials of Knowledge Management. Wiley, New Jersey (2003)

Breuker, J., Casanovas, P., Klien, M.C.A., Francesconi, E.: Law, Ontologies and Semantic WEB. IOS Press, Amsterdam (2009)

Hauke, K., Kutzner, I., Marciniak, K., Owoc, M.: Creation of the urban knowledge portal: e-learning and knowledge inventor context. In: The 11th International Conference on Semantic, Grid and Knowledge, SKG 2015 Beijing, China

Tiwana, A.: Knowledge Management Toolkit. PTR, Upper Saddle River (1999)

Townsend, A.M.: Smart Cities: Big Data, Civic Hackers, and the Quest for a New Utopia. W. W. Norton & Company, New York (2014)

Berlin Haupstadt. http://www.berlin.de/

Urban Knowledge, May 2015. http://www.igi-global.com/chapter/knowledge-creation-urban-knowledge-environment/25482

London City. http://www.london.gov.uk/

Wroclaw miasto. http://www.wroclaw.pl/

Chourabi, H., Nam, T., Walker, S., Gil-Garcia, J.R., Mellouli, S., Nahon, K., Pardo, T.A., Scholl, H.J.: Understanding smart cities: An integrative framework. In: Proceedings of the Annual Hawaii International Conference on System Sciences, Maui, HI, USA, pp. 2289–2297 (2012)

Conceptual Navigation for Polyadic Formal Concept Analysis

Sebastian Rudolph[1], Christian Săcărea[2], and Diana Troancă[2(✉)]

[1] Technische Universität Dresden, Dresden, Germany
sebastian.rudolph@tu-dresden.de
[2] Babeş-Bolyai, Cluj-Napoca, Romania
{csacarea,dianat}@cs.ubbcluj.ro

Abstract. Formal Concept Analysis (FCA) is a mathematically inspired field of knowledge representation with wide applications in knowledge discovery and decision support. Polyadic FCA is a generalization of classical FCA that instead of a binary uses an arbitrary, n-ary incidence relation to define *formal concepts*, i.e., data clusters in which all elements are interrelated. We discuss a paradigm for navigating the space of such (formal) concepts, based on so-called membership constraints. We present an implementation for the cases $n \in \{2, 3, 4\}$ using an encoding into answer-set programming (ASP) allowing us to exploit highly efficient strategies offered by optimized ASP solvers. For the case $n = 3$, we compare this implementation to a second strategy that uses exhaustive search in the concept set, which is precomputed by an existing tool. We evaluate the implementation strategies in terms of performance. Finally, we discuss the limitations of each approach and the possibility of generalizations to n-ary datasets.

1 Introduction

Conceptual knowledge is closely related to a deeper understanding of existing facts and relationships, but also to the process of arguing and communicating why something happens in a particular way. Formal Concept Analysis (FCA) [6] is a mathematical theory introduced by Wille, being the core of Conceptual Knowledge Processing [24]. It emerged from applied mathematics and quickly developed into a powerful framework for knowledge representation. It is based on a set-theoretical semantics and provides a rich amount of mathematical instruments for the representation, acquisition, retrieval, discovery and further processing of knowledge.

FCA defines concepts as maximal clusters of data in which all elements are mutually interrelated. In FCA, data is represented in a basic data structure, called *formal context*. A *dyadic* formal context consists of two sets, one of *objects* and another of *attributes* and a binary relation between them, expressing which objects have which attributes. From such dyadic formal contexts, *formal concepts* can be extracted using concept forming operators, obtaining a mathematical structure called *concept lattice*. Thereby, the entire information contained in

E. Mercier-Laurent and D. Boulanger (Eds.): AI4KM 2016, IFIP AICT 518, pp. 50–70, 2018.
https://doi.org/10.1007/978-3-319-92928-6_4

the formal context is preserved. The concept lattice and its graphical representation as an order diagram can then serve as the basis for communication and further data analysis. Navigation in concept lattices enables exploring, searching, recognizing, identifying, analyzing, and investigating; this exemplifies the fruitfulness of this approach for knowledge management.

In subsequent work, Lehmann and Wille extended dyadic FCA to a triadic setting (3FCA) [16], where *objects* are related to *attributes* under certain *conditions*. The *triadic concepts* (short: *triconcepts*) arising from such data, can be arranged in mathematical structures called *trilattices*. Trilattices can be, up to some conditions, graphically represented as a triangular diagram, yet, this kind of knowledge representation is much less useful and intuitive than its dyadic counterpart, because of the difficulties of reading and navigating in such triadic diagrams. Even if the theoretical foundations of trilattices and that of 3FCA have been intensely studied, there is still a need for a valuable navigation paradigm in triadic concept sets. To overcome these difficulties, we proposed in 2015 a navigation method for triadic conceptual landscapes based on a neighborhood notion arising from dyadic concept lattices obtained by projecting along a dimension [20]. This method enables exploration and navigation in triconcept sets by locally displaying a smaller part of the space of triconcepts, instead of displaying all of them at once.

Voutsadakis [22] further generalized the idea from dyadic and triadic to n-adic data sets, introducing the term *Polyadic Concept Analysis*. He describes concept forming operators in the n-dimensional setting as well as the basic theorem of polyadic concept analysis, a generalization of earlier results by Wille [23].

FCA was successfully used on triadic or tetradic ($n = 4$) datasets such as folksonomies [14], data logs of rental services [2] or data about mobile operators [12]. However, a common problem for n-ary concept sets is their size and complexity. Even for $n = 2$ and for relatively small data sets, the number of formal concepts tends to be quite large (it can be of exponential size in the worst case), which makes the graphical representation of these sets in their entirety unusable for practical purposes. Several strategies have been proposed to overcome this problem. For instance, Dragoş et al. [4,5] are using a circular representation for triadic data while investigating users' behavioral patterns in e-learning environments. Săcărea [21] uses a graph theoretical approach to represent triadic concept sets obtained from medical data. For n-adic concept sets with $n \geq 4$, no navigation strategies have been presented yet.

In 2015, we introduced membership constraints for n-adic concepts in order to narrow down the set of user-relevant n-concepts and to focus on a certain data subset one is interested to explore or start exploration from [19]. As opposed to classical navigation tools, conceptual navigation has at its core a formal concept, i.e. a complete cluster of knowledge. We discussed the problem of satisfiability of membership constraints, determining if a formal concept exists whose object and attribute sets include certain elements and exclude others.

In the current paper, we consider a general navigation paradigm for the space of polyadic concepts and implement this paradigm for the dyadic, triadic and tetradic case. For the triadic case, we try two different implementations. The first one uses the capabilities of Answer Set Programming (ASP) for computing concepts and solving the corresponding membership constraint satisfaction problem. By using this strategy, the implementation also explores optimization strategies offered by ASP. The second strategy is based on an exhaustive search of the set of polyadic concepts. The concept set is no longer computed using the ASP encoding but by one of the existing 3FCA tools. Finally, we evaluate the performance of these two strategies in terms of implementation and computation speed and we discuss the limitations of each approach and show that the ASP approach can be extended to any n-ary dataset.

2   Preliminaries

2.1   Formal Concept Analysis

In this section, we briefly present the necessary basic notions and definitions. For a more detailed introduction, please refer to [6].

A dyadic context is defined to be a triple (G, M, I), where G and M are sets and $I \subseteq G \times M$ is a binary relation. The set G is called *set of objects*, M is the *set of attributes* and I is the *incidence relation*. Concepts are extracted herefrom using *derivation operators*. For $A \subseteq G$ we define $A' = \{m \in M \mid \forall g \in A, gIm\}$, i.e. A' consists of all common attributes of all object from A. Dually, for a set $B \subseteq M$ of attributes, we define B' to be the set of objects sharing all attributes from B. The derivation operators are a Galois connection on the power sets of G and M, respectively. A *formal concept* is a pair (A, B) of objects and attributes, respectively, with $A' = B, B' = A$. The set of all formal concepts is ordered by the subconcept-superconcept relationship: If $(A, B), (C, D)$ are concepts, then $(A, B) \leq (C, D)$ if and only if $A \subseteq C$ (which is equivalent to the condition $D \subseteq B$). By this, the set of all concepts becomes a complete lattice, called *concept lattice*.

The triadic case has been studied by Lehmann and Wille ([16]). The fundamental structures used by triadic FCA are those of a triadic formal context and a triadic formal concept, also referred to as triadic concept or short triconcept. For a better understanding, we define the triadic context first, and then explain the notion of formal concept for the general case.

Definition 1. *A triadic formal context* $\mathbb{K} = (K_1, K_2, K_3, Y)$ *is defined as a quadruple consisting of three sets and a ternary relation* $Y \subseteq K_1 \times K_2 \times K_3$. K_1 *represents the set of objects,* K_2 *the set of attributes and* K_3 *the set of conditions. The notation for an element of the incidence relation is* $(g, m, b) \in Y$ *or* $b(g, m)$ *and it is read object g has attribute m under condition b.*

Polyadic FCA is a direct generalization of the dyadic or triadic case, where n (not necessarily different) non-empty sets are related via an n-ary relation. An n-concept is a maximal cluster of n sets, with every element being interrelated with all the others.

Definition 2. *Let $n \geq 2$ be a natural number. An n-context is an $(n+1)$-tuple $\mathbb{K} := (K_1, K_2, \ldots, K_n, Y)$, where $K_1, K_2, \ldots, K_n$ are sets and Y is an n-ary relation $Y \subseteq K_1 \times K_2 \times \cdots \times K_n$.*

Definition 3. *The n-concepts of an n-context $(K_1, \ldots, K_n, Y)$ are exactly the n-tuples $(A_1, \ldots, A_n)$ that satisfy $A_1 \times \cdots \times A_n \subseteq Y$ and which are maximal with respect to component-wise set inclusion. $(A_1, \ldots, A_n)$ is called a* proper *n-concept if $A_1, \ldots, A_n$ are all non-empty.*

Example 1. Finite dyadic contexts can be represented as cross-tables, rows being labeled with object names, columns with attribute names. In the triadic case, objects are related to attributes and conditions via a ternary relation and the corresponding triadic context can be thought of as a 3D cuboid, the ternary relation being marked by filled cells. Triadic contexts are usually unfolded into a series of dyadic "slices", like in the following example, where we consider a triadic context (K_1, K_2, K_3, Y) where the object set K_1 consists of authors of scientific papers, the attribute set K_2 contains conference names/journal names while the conditions K_3 are the publication years. For this small selection we obtain a $2 \times 4 \times 2$ triadic context, the "slices" being labeled by condition names (Fig. 1).

2014	Corr	ICC	PIMRC	HICSS
Rumpe	×			
Alouni	×	×	×	

2015	Corr	ICC	PIMRC	HICSS
Rumpe	×			×
Alouni	×	×		

Fig. 1. DBLP data: author, conference/journal, year

There are exactly six triconcepts of this context, i.e., maximal 3D cuboids full of incidences:

- $(\{Rumpe, Alouni\}, \{Corr\}, \{2014, 2015\})$,
- $(\{Alouni\}, \{Corr, ICC, PIMRC\}, \{2014\})$,
- $(\{Alouni\}, \{Corr, ICC\}, \{2014, 2015\})$,
- $(\{Rumpe\}, \{Corr, HICSS\}, \{2015\})$,
- $(\emptyset, \{Corr, ICC, PIMRC, HICSS\}, \{2014, 2015\})$ and
- $(\{Rumpe, Alouni\}, \{Corr, ICC, PIMRC, HICSS\}, \emptyset)$.

The first four of these triconcepts are called *proper*.

If $\mathbb{K} = (K_1, \ldots, K_n, Y)$ is an n-context, *membership constraints* are indicating restricting conditions by specifying which specific elements $a_j \in K_j$ must be included in the jth component of an n-concept, respectively which elements $b_j \in K_j$, $j = 1, \ldots, n$ must be excluded therefrom. We investigated the question of satisfiability of such membership constraints, i.e., to determine if there are any formal n-concepts which are satisfying the inclusion and exclusion requirements [19].

Definition 4. *An n-adic membership constraint on an n-context $\mathbb{K} = (K_1, \ldots, K_n, R)$ is a $2n$-tuple $\mathbb{C} = (K_1^+, K_1^-, \ldots, K_n^+, K_n^-)$ with $K_i^+ \subseteq K_i$ called required sets and $K_i^- \subseteq K_i$ called forbidden sets.*

An n-concept $(A_1, \ldots, A_n)$ of $\mathbb{K}$ is said to satisfy *such a membership constraint if $K_i^+ \subseteq A_i$ and $K_i^- \cap A_i = \emptyset$ hold for all $i \in \{1, \ldots, n\}$.*

We let $\mathrm{Mod}(\mathbb{K}, \mathbb{C})$ $(\mathrm{Mod}_p(\mathbb{K}, \mathbb{C}))$ denote the set of all (proper) n-concepts of $\mathbb{K}$ that satisfy $\mathbb{C}$.

An n-adic membership constraint $\mathbb{C}$ is said to be (properly) satisfiable with respect to $\mathbb{K}$, if it is satisfied by one of its (proper) n-concepts, that is, if $\mathrm{Mod}(\mathbb{K}, \mathbb{C}) \neq \emptyset$ $(\mathrm{Mod}_p(\mathbb{K}, \mathbb{C}) \neq \emptyset)$.

We have shown that the problem of deciding satisfiability of a membership constraint w.r.t. an n-context is NP-complete in general [19]. The intractability easily carries over to proper satisfiability.

2.2 Answer Set Programming

Answer Set Programming (ASP) [7] is a logic programming formalism and hence uses a declarative approach to solve NP-hard problems. As opposed to the imperative approach, where the programmer tells the computer what steps to follow in order to solve a problem, declarative approaches merely describe the problem while the process of solving is delegated to generic strategies applied by highly optimized ASP engines. Mainly, in ASP one has to express the problem in a logic programming format consisting of facts and rules, so that the solutions of the problem correspond to models of the logic program.

In what follows, we briefly introduce the syntax and semantics of normal logic programs under the stable model semantics [7,10]. We will present directly the syntax used in the source code in order to avoid a translation phase from one syntax to the other.

Let $\mathcal{D}$ denote the *domain*, i.e. a countable set of elements, also called *constants*. Next, we define an *atom* as an expression of the type $\mathrm{p}(t_1, \ldots, t_n)$, where p is a *predicate* of arity $n \geq 0$ and every t_i is an element from the domain or a variable, denoted by an upper case letter. An atom is called *ground* if it is variable-free. The set of all ground atoms over $\mathcal{D}$ is denoted by $\mathcal{G}_\mathcal{D}$. A *(normal)* *rule* ρ is defined as:

$$a_1 \leftarrow b_1 \wedge \ldots \wedge b_k \wedge \sim b_{k+1} \wedge \ldots \wedge \sim b_m,$$

where $a_1, b_1, \ldots, b_m$ are atoms $m \geq k \geq 0$ with the observation that the left or the right part of the rule might be missing, but not both at the same time. The left part of the rule, i.e. the part before "$\leftarrow$" is called *head*, denoted $H(\rho) = \{a_1\}$, while the right part is the *body* of the rule, denoted $B(\rho) = \{b_1, \ldots b_k, \sim b_{k+1}, \ldots, \sim b_m\}$. As mentioned previously, a rule does not necessarily contain a non-empty head and body, namely, when the head of the rule is empty, we call the rule a *constraint* and, when the body of the rule is missing and a_1 is ground, it is called a *fact*. In case the rule is a fact, we usually omit the sign "$\leftarrow$" and write just the atom in the head of the rule. In the definition of the rule, "$\sim$" denotes default negation, which refers to the absence of information as opposed to classical negation "$\neg$" which implies that the negated information is present. Intuitively, "$\sim a$" means that $a \notin I$, while $\neg a$ implies $\neg a \in I$, for an interpretation I, where $I \subseteq \mathcal{G}_D$ can be understood as the set of ground atoms which are true. Furthermore, we denote $B^+(\rho) = \{b_1, \ldots, b_k\}$ and $B^-(\rho) = \{b_{k+1}, \ldots, b_m\}$. A rule ρ is called *safe* if each variable in ρ occurs in $B^+(\rho)$. Finally, we define a propositional normal logic program as a finite set of normal rules.

In order to define when a program Π is satisfied by an interpretation I, let $\mathcal{U}_\Pi$ denote the subset of constants from the domain that appear in the program Π and $Gr(\Pi)$ the grounded program, i.e. the set of grounded rules obtained by applying all the possible substitutions from the variables to the constants in $\mathcal{U}_\Pi$, for all the rules $\rho \in \Pi$. We say that an interpretation $I \subseteq \mathcal{G}_D$ satisfies a normal ground rule $\rho \in \Pi$ if and only if the following implication holds:

$$B^+(\rho) \subseteq I, B^-(\rho) \cap I = \emptyset \Rightarrow H(\rho) \subseteq I.$$

Then the interpretation I satisfies a non-ground rule if it satisfies all the possible groundings of the rule. Finally, the interpretation I satisfies a program Π if it satisfies all its rules, i.e. it satisfies the grounded program $Gr(\Pi)$.

An interpretation $I \in \mathcal{G}_D$ is called an *answer set* or a *stable model* [10] of the program Π if and only if it is the $\subseteq$-minimal model satisfying the reduct Π^I defined by $\Pi^I = \{H(\rho) : -B^+(\rho) \mid I \cap B^-(\rho), \rho \in Gr(\Pi)\}$. Furthermore, we define the set of *cautiously entailed* ground atoms as the intersection of all answer sets.

ASP solving is split in two phases. In the first phase, a grounder has to be used in order to process the logic program into a finite variable-free propositional representation of the problem encoding. In the next phase, a solver uses as input the output of the grounder and computes the solutions, i.e. the answer sets of the problem.

Researchers from the University of Potsdam developed a tool suite called Potassco[1], which is a collection of answer set solving software. In our research we used the solving tools from the this collection [8], since it is currently the most prominent solver leading in the latest competitions [1]. In what follows, we will describe these tools in more details.

The first tool, Gringo is a grounder that can be used in the first phase in order to transform the initial encoding into an equivalent variable-free, i.e. ground, program. Gringo has a simple, but comprehensive syntax that can express different types of rules (normal, choice, cardinality, etc.), constraints (integrity, cardinality, etc.), but also optimization statements. Intuitively, a constraint expresses a

---

[1] http://potassco.sourceforge.net/.

"forbidden" behavior of the models, i.e. if the body of the constraint is true then the model is not a stable model. The output of Gringo is in the *smodels* format, which is an intermediate format used for the ASP solver input.

The second tool, Clasp, is a solver that uses the smodels format for the input and computes the answer sets of the program. The output of Clasp can be configured by the user and shows all or some of the following details: the number of solutions and whether the problem is *satisfiable* (i.e. if it has at least one stable model) or *unsatisfiable*, the solutions, and the detailed time of the execution. The format of the answer sets is also configurable by specifying in the encoding which predicates shall be printed in the output.

Both tools, Gringo and Clasp, were combined into one tool, called Clingo, in order to avoid processing the ASP program with Gringo and then further process the output with Clasp. Avoiding the intermediate step is particularly useful if the grounded program is not of interest and the only results needed are the answer sets of the program. Furthermore, in case one needs to compute the execution time of the whole ASP solving process, the integrated tool shows the cumulative duration of the two phases, grounding and solving.

3   Encoding Membership Constraints in Answer Set Programming

Given that satisfiability of membership constraints can in general be NP-complete, it is nontrivial to find efficient algorithms for computing membership-constraint-satisfying concepts. We note here that the problem can be nicely expressed in answer set programming using an encoding introduced in a previous paper [19]. We will consider the n-adic case. Given an n-adic membership constraint $\mathbb{C} = (K_1^+, K_1^-, \ldots, K_n^+, K_n^-)$ on an n-context $\mathbb{K} = (K_1, \ldots, K_n, R)$, we represent the specific problem by the following set of ground facts $F_{\mathbb{K},\mathbb{C}}$:

- $\mathtt{set}_i(a)$ for all $a \in K_i$,
- $\mathtt{rel}(a_1, \ldots, a_n)$ for all $(a_1, \ldots, a_n) \in R$,
- $\mathtt{required}_i(a)$ for all $a \in K_i^+$, and
- $\mathtt{forbidden}_i(a)$ for all $a \in K_i^-$.

Now we let P denote the following fixed answer set program (with rules for every $i \in \{1, \ldots, n\}$):

$$\mathtt{in}_i(x) \leftarrow \mathtt{set}_i(x) \wedge \sim\!\mathtt{out}_i(x)$$
$$\mathtt{out}_i(x) \leftarrow \mathtt{set}_i(x) \wedge \sim\!\mathtt{in}_i(x)$$
$$\leftarrow \bigwedge\nolimits_{j \in \{1, \ldots, n\}} \mathtt{in}_j(x_j) \wedge \sim\!\mathtt{rel}(x_1, \ldots, x_n)$$
$$\mathtt{exc}_i(x_i) \leftarrow \bigwedge\nolimits_{j \in \{1, \ldots, n\} \setminus \{i\}} \mathtt{in}_j(x_j) \wedge \sim\!\mathtt{rel}(x_1, \ldots, x_n)$$
$$\leftarrow \mathtt{out}_i(x) \wedge \sim\!\mathtt{exc}_i(x)$$
$$\leftarrow \mathtt{out}_i(x) \wedge \mathtt{required}_i(x)$$
$$\leftarrow \mathtt{in}_i(x) \wedge \mathtt{forbidden}_i(x)$$

Intuitively, the first two lines "guess" an n-concept candidate by stipulating for each element of each K_i if they are in or out. The third rule eliminates a candidate if it violates the condition $A_1 \times \ldots \times A_n \subseteq R$, while the fourth and fifth rule ensure the maximality condition for n-concepts. Finally, the sixth and the seventh rule eliminate n-concepts violating the given membership constraint.

There is a one-to-one correspondence between the answer sets X of $F_{\mathbb{K},\mathbb{C}} \cup P$ and the n-concepts of $\mathbb{K}$ satisfying $\mathbb{C}$ obtained as $(\{a \mid \mathrm{in}_1(a) \in X\}, \ldots, \{a \mid \mathrm{in}_n(a) \in X\})$. Consequently, optimized off-the-shelf ASP tools can be used for checking satisfiability but also for enumerating all satisfying n-concepts.

4   Navigation

In this section, we describe a strategy for navigating the space of proper n-concepts of an n-context.[2] The basic idea is to use intuitive representations of "subspaces" of the overall space by specifying which elements must be included in or excluded from a certain proper n-concept component A_i. Obviously, such a subspace is identified by a membership constraint $\mathbb{C} = (K_1^+, K_1^-, \ldots, K_n^+, K_n^-)$ specifying exactly the included and excluded elements for each component of the n-concepts. The n-concepts in the "subspace" associated with $\mathbb{C}$ are then the n-concepts from $\mathrm{Mod}_p(\mathbb{K}, \mathbb{C})$. Visually, $\mathbb{C}$ can be represented by displaying $K_1, \ldots, K_n$ as n lists and indicating for every element if it is included, excluded, or none of the two (undetermined). The user can then choose to restrict the space further by indicating for an undetermined element of some K_i, if it should be included or excluded. What should, however be avoided is that by doing so, the user arrives at an empty "subspace", i.e., a membership constraint that is not satisfied by any proper n-concept (i.e., $\mathrm{Mod}_p(\mathbb{K}, \mathbb{C}) = \emptyset$). To this end, we will update the membership constraint directly after the user interaction in order to reflect all necessary inclusions and exclusions automatically following from the user's choice. Assume $\mathbb{C} = (K_1^+, K_1^-, \ldots, K_n^+, K_n^-)$ is the membership constraint after the user interaction. The updated constraint can be described by $\mathbb{C}' = (L_1^+, L_1^-, \ldots, L_n^+, L_n^-)$, where

$$L_i^+ = \bigcap_{(A_1, \ldots, A_n) \in \mathrm{Mod}_p(\mathbb{K}, \mathbb{C})} A_i$$

and

$$L_i^- = \bigcap_{(A_1, \ldots, A_n) \in \mathrm{Mod}_p(\mathbb{K}, \mathbb{C})} K_i \setminus A_i.$$

It is then clear that after such an update, for every element e of some K_i which is still undetermined by $\mathbb{C}'$, there exist proper n-concepts $(E_1, \ldots, E_n)$ and $(F_1, \ldots, F_n)$ in $\mathrm{Mod}_p(\mathbb{K}, \mathbb{C}')$ with $e \in E_i$ but $e \notin F_i$. Consequently, whatever undetermined element the user chooses to include or exclude, the resulting

---

[2] Non-proper concepts are considered out of scope for knowledge exploration, thus we exclude them from our consideration. The described navigation would, however, also work if these concepts were taken into account.

membership constraint will be properly satisfiable. If the updated constraint $\mathbb{C}' = (L_1^+, L_1^-, \ldots, L_n^+, L_n^-)$ determines for every element if it is included or excluded (i.e., if $L_i^+ \cup L_i^- = K_i$ holds for every i), the user's navigation has narrowed down the space to the one proper n-concept $(L_1^+, \ldots, L_n^+)$.

Considering the example from the previous section, assume the user has specified the inclusion of the attribute $Corr$ in K_2 and the exclusion of the attribute ICC from K_2, i.e.,

$$\mathbb{C} = (\emptyset, \emptyset, \{Corr\}, \{ICC\}, \emptyset, \emptyset).$$

The proper 3-concepts of $\mathbb{K}$ satisfying $\mathbb{C}$ are

$$C_1 = (\{Rumpe, Alouni\}, \{Corr\}, \{2014, 2015\}) \quad \text{and}$$
$$C_2 = (\{Rumpe\}, \{Corr, HICSS\}, \{2015\}),$$

therefore, we would obtain the updated constraint

$$\mathbb{C}' = (\{Rumpe\}, \emptyset, \{Corr\}, \{ICC, PIMRC\}, \{2015\}, \emptyset).$$

If the user now decided to additionally exclude 2014 from K_3, leading to the constraint

$$\mathbb{C}'' = (\{Rumpe\}, \emptyset, \{Corr\}, \{ICC, PIMRC\}, \{2015\}, \{2014\}),$$

the only proper 3-concept satisfying it is C_2. Consequently, $\mathbb{C}''$ will be updated to

$$\mathbb{C}''' = (\{Rumpe\}, \{Alouni\}, \{Corr, HICSS\}, \{ICC, PIMRC\}, \{2015\}, \{2014\}),$$

which then represents the final state of the navigation.

5   Implementation

Following the general scheme described in the previous section, we implemented a navigation tool for the cases $n \in \{2, 3, 4\}$, using different strategies for $n = 3$. The two fundamentally different approaches differ in the method of computing the concepts (ASP vs different tool), as well as in which navigation step the concepts are computed.

In 2015, we proposed the ASP encoding for the membership constraint satisfiability problem presented in Sect. 3 and discussed how it could be deployed in an interactive search scenario [19]. For the first approach presented in this paper[3], we extended and implemented this scenario using diverse ASP optimization techniques. For grounding and solving in the ASP navigation tool, we used Clingo from the Potassco collection [8] for the reasons mentioned in Sect. 2.2.

Recall that ASP solves a search problem by computing answer sets, which represent the models of a given answer set program (the so-called stable models)

---

[3] https://sourceforge.net/projects/asp-concept-navigation.

[7,9,11,17,18]. Our encoding, as presented in Sect. 3, is such that given $\mathbb{K}$ and $\mathbb{C}$, an answer set program is created, such that there is a one-to-one correspondence between the answer sets and the n-concepts of $\mathbb{K}$ satisfying $\mathbb{C}$.

The known facts in a membership constraint satisfiability problem are the elements of the context $K_i, i \in \{1, \ldots, n\}$, the n-adic relation Y and the sets of required and forbidden elements. The answer set program can be conceived as a declarative implementation of the following "guess & check" strategy:

- start from an empty constraint $\mathbb{C}$
- decide for each element $a \in K_i, i \in \{1, \ldots, n\}$, if $a \in K_i^+$, i.e. included, or $a \in K_i^-$, i.e. excluded, hence reaching a membership constraint of the form $\mathbb{C} = (K_1^+, K_1^-, \ldots, K_n^+, K_n^-)$ with $K_i^+ \cup K_i^- = K_i$ for every i
- check if $(K_1^+, \ldots, K_n^+)$ is component-wise maximal w.r.t. $K_1^+ \times K_2^+ \times \ldots \times K_n^+ \subseteq Y$
- check if the required and forbidden elements are assigned correspondingly in the obtained membership constraint C, i.e. required elements belong to K_i^+, forbidden elements belong to K_j^-, for some $i, j \in \{1, \ldots, n\}$.

At any step, if one of the conditions is violated, the membership constraint is eliminated from the set of models. Hence, in the end we obtain all the membership constraints that correspond to n-concepts satisfying the given restrictions. The ASP encoding can be easily extended to retrieve only the proper n-concepts satisfying $\mathbb{C}$, by adding an additional check that ensures $|K_i^+| > 0$ for every i.

Recall that the *cautious* option of ASP iteratively computes the intersection over all answer sets, in the order in which they are computed by the ASP solver. However, regardless of the ASP solver, the last outputted solution when using the *cautious* option is always the intersection of all answer sets of the program. Later in this section, we show how this option can be utilized to optimize the propagation phase of the navigation and, in the evaluation section, we present statistics showing that it has a great impact on the execution time of the tool we developed.

The propagation algorithm tests for all elements that are still in an undetermined state, which of the possible decisions on that element (*in* or *out*) give rise to a satisfiable answer set program. In case one of the decisions generates an unsatisfiable problem, the complementary choice is automatically made. Remember that, as discussed in the previous section, when starting from a satisfiable setting, it cannot be the case that both choices generate an unsatisfiable program.

The alternative to explicitly testing all the possible choices for every element in an undetermined state is to compute all the answer sets for the already added constraints and to obtain their intersection. This intersection contains the *in* and *out* choices that need to be propagated, since their complementary constraints are not included in any answer set and hence, would generate an unsatisfiable program. This approach is formally described in Algorithm 1.

Algorithm 1. propagation of user decisions optimized

---

function PROPAGATEOPTIMIZED($\mathbb{K}, \mathbb{C}$)
Input: n-context $\mathbb{K}$, membership constraint $\mathbb{C}$
Output: updated membership constraint
Data: membership constraint $\mathbb{C} = (K_1^+, K_1^-, \ldots, K_n^+, K_n^-)$

   for all $i \in \{1, \ldots, n\}$ **do**
$$L_i^+ = \bigcap_{(A_1, \ldots, A_n) \in \mathrm{Mod}_p(\mathbb{K}, \mathbb{C})} A_i$$
$$L_i^- = \bigcap_{(A_1, \ldots, A_n) \in \mathrm{Mod}_p(\mathbb{K}, \mathbb{C})} K_i \setminus A_i.$$
   end for
   $\mathbb{C} = (L_1^+, L_1^-, \ldots, L_n^+, L_n^-)$
   return $\mathbb{C}$
end function

Algorithm 1 was implemented in the ASP navigation tool using the *cautious* option described previously. The implementation requires a single call to the ASP solver which computes the intersection of the models in the answer set. This intersection actually corresponds to the membership constraint containing all the inclusions and exclusions that need to be propagated. In comparison, for the simple propagation algorithm, multiple calls to the ASP solver are necessary: For each element that is in an undetermined state, two membership constraint satisfiability problems are generated, checking whether adding the element to the required objects, respectively to the forbidden objects, generates an unsatisfiable program. The cautious-based optimized propagation algorithm proved to drastically decrease the computation time as well as the memory usage, hence improving the performance of the interactive navigation tool. The experimental results are described in more detail in the evaluation section.

The optimized ASP approach was implemented and evaluated for triadic data. Furthermore, to show that it is easily extended to any n-adic context, we also implemented the dyadic and tetradic case, however, without evaluating their performance on real data sets. In fact, the only modifications that need to be made when updating the context's dimension are to add the new sets to the ASP encoding and to update the graphical interface and the context loader to the new context dimension.

The second approach for the navigation is a brute force implementation[4] and uses an exhaustive search in the whole space of n-concepts. Hence, a prerequisite for this tool is to previously compute all n-concepts using an existing tool. We implemented the triadic case ($n = 3$) and used Trias [13] to compute the triadic formal concepts. For that reason, the input for the navigation tool is adapted to Trias' output format.

This approach follows the same steps described in Algorithm 1, however it uses different methods for implementing them. The first main difference lies in

---

[4] https://sourceforge.net/projects/brute-force-concept-navigation.

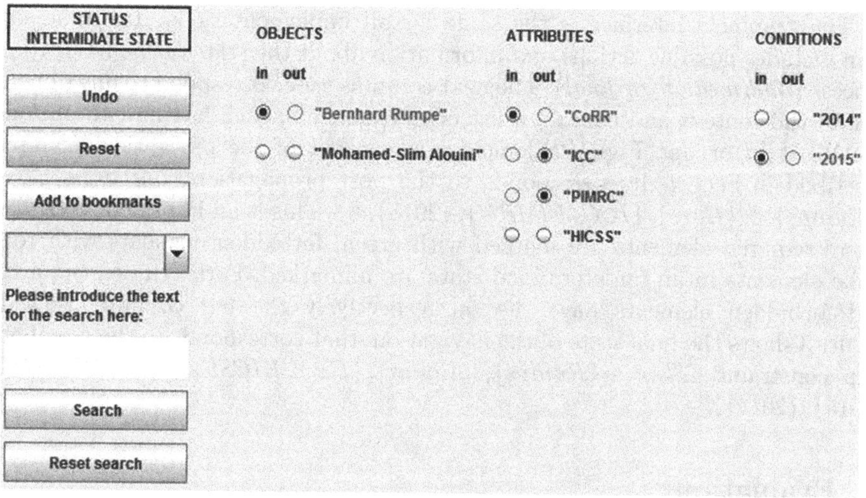

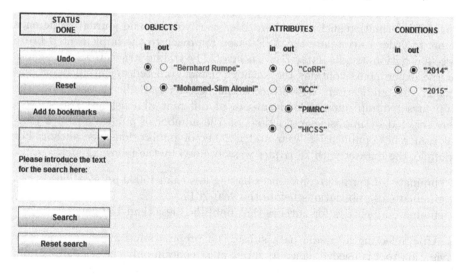

Fig. 2. Screenshot navigation tool: intermediate state (Color figure online)

Fig. 3. Screenshot navigation tool: final state

the method of computing the triconcepts. Instead of computing them at each step using a declarative approach, in the brute force approach, all triconcepts are computed in the preprocessing phase using an existing algorithm. In a navigation step an exhaustive search is necessary in order to select the subset of triconcepts that satisfy the constraints and compute the intersection. This subset of triconcepts is successively pruned in each navigation step until it contains a single triconcept, which represents the final state of the navigation.

The graphical interface is the same for all implementations. The first column includes possible actions and information about the state of the navigation process (*intermediate* or *final*). The next columns each correspond to one dimension of the context and contain a list of the elements, each having two options next to it *in* or *out*. Figure 2 depicts a screenshot of the navigation example described in Sect. 4. It corresponds to the post propagation constraint $\mathbb{C}' = (\{Rumpe\}, \emptyset, \{Corr\}, \{ICC, PIMRC\}, \{2015\}, \emptyset)$. This is an intermediate state, where required elements are marked with green, forbidden elements with red, while elements in an undetermined state are unmarked. Furthermore, required and forbidden elements have the *in*, respectively the *out* column checked. Figure 3 shows the final state of the navigation, that corresponds to the membership constraint $\mathbb{C}''' = (\{Rumpe\}, \{Alouni\}, \{Corr, HICSS\}, \{ICC, PIMRC\}, \{2015\}, \{2014\})$.

6 Evaluation

In order to evaluate the implemented tools, we ran experiments on the dblp database[5]. The dblp database indexes conference and journal publications and contains information such as author, title, year, volume, and journal/conference name. In order to compare the ASP-based approach to the implemented brute force navigation, triadic datasets are needed. The triadic structure that we chose for the experiments contains the author's name, conference/journal name and year of the publication. We extracted the described triadic dataset from the dblp mysql dump and selected subsets of different dimensions. The subsets were selected by imposing restrictions on the number of publications per journal/conference, publication year and number of publications per author. For example, the dataset with 28 triples was obtained by the following steps:

– eliminate all journals/conferences having less than 15000 publications
– eliminate all publications before the year 2014
– eliminate all entries for authors that published less than 150 papers

After selecting a triadic data subset, no preprocessing phase for the ASP navigation tool is needed, since its input must contain only the triadic relation. However, the brute force navigation tool requires a preprocessing phase. First the triconcept set needs to be computed with the Trias algorithm, hence the Trias tool[6] needs to be installed separately. If using the Trias algorithm without a database connection, the standard input file requires numeric data. Hence, in order to format the data according to the Trias tool input format, the elements of the dataset need to be indexed. After running Trias to obtain the triconcepts, the output needs to be formatted again before using the brute force navigation tool. Mainly the dimensions and encodings of the object, attribute and condition sets need to be added, so that the navigation tool can output the names of

---

[5] http://dblp.uni-trier.de/.
[6] https://github.com/rjoberon/trias-algorithm.

Table 1. ASP navigator experiments

Number of objects	Number of attributes	Number of conditions	Number of triples	ASP navigation data loading time (s)	ASP navigation average step time (s)
2	15	2	28	0.015	0.1873
14	62	5	680	0.109	0.2315
41	67	7	2514	0.374	0.3278
68	67	8	4478	0.546	0.5930
83	67	9	5987	0.660	0.6350
108	67	10	8133	1.070	1.1940

Table 2. Brute force navigator experiments

Number of objects	Number of attributes	Number of conditions	Number of triples	Trias pre-processing time (s)	Brute force data loading time (s)	Brute force average step time (s)
2	15	2	28	0.27	0.016	0.0060
14	62	5	680	1.04	0.421	0.0047
41	67	7	2514	23.24	1.950	0.0219
68	67	8	4478	644.758	4.384	0.0530
83	67	9	5987	2152.839	6.992	0.1600
108	67	10	8133	>2h		

the elements and not their indexes. Only after these preprocessing steps can a user interactively navigate in the tricontext using the brute force navigation tool. Obviously, different formats for the input of the navigation tool can be implemented, but for the purpose of comparing the two tools we implemented one single input format based on the standard Trias output.

For measuring the runtimes of the two navigation tools, we have evaluated their performance on six different datasets containing between 28 and 8133 triples. The datasets are described in Tables 1 and 2 (we ran both tools on the same datasets), where objects are identified with author names, attributes with conferences/journal names and conditions with the publication years. For each dataset, we chose some random navigation paths through the data, which contain between 4 and 13 navigation steps and end when a final state, i.e., a triconcept, is reached. By navigation step we understand not only the action of a user choosing an element as *in* or *out*, but also the subsequent propagation phase. In order to compare the two approaches we computed the average navigation step time for each dataset and measured the time used for loading the data. This information can be obtained from the file *statistics.log* which is created as an output by the navigation tools. Furthermore, for the brute force navigation we also measured the preprocessing time, i.e. the time that Trias needs to compute the triconcepts. Note that the time needed to index the dataset for the Trias

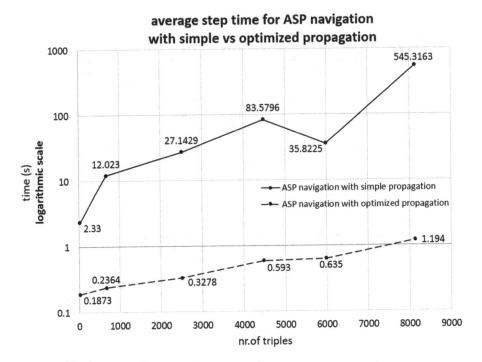

Fig. 4. Average step time for ASP navigation with simple propagation vs optimized propagation

input, as well as to add the encodings to the Trias output to obtain the input for the navigation tool, were excluded from this analysis, since this processing phase can be avoided by implementing different input/output formats for the Trias tool or for the brute force navigation tool. We denote the data loading time plus the preprocessing time as offline time. In case of the ASP navigation tool, the offline time equals the data loading time, since no preprocessing is needed. The experiments were run on an Intel(R) Core(TM) I7-3630QM CPU @ 2.40 GHz machine with 4 GB RAM and 6 M Cache.

First, we compared the different propagation implementations for the ASP approach: simple propagation vs. optimized propagation. The results are shown in Fig. 4, where the y-axis depicts the logarithmically scaled time of execution, while the x-axis corresponds to the size of the relation. Besides the big difference in the execution time of each step, the ASP navigation tool with simple propagation uses a lot of memory. For the context with 8133 triples after a few navigation steps the execution was stopped by the system because it reached the limit memory of 4 GB RAM. In comparison, this problem does not occur for the navigation tool with optimized propagation.

Next, we ran experiments on the same dataset to compare the ASP navigation tool with optimized propagation to the brute force navigation tool. Figure 5 shows the offline time of the ASP navigation tool vs. the brute force navigation

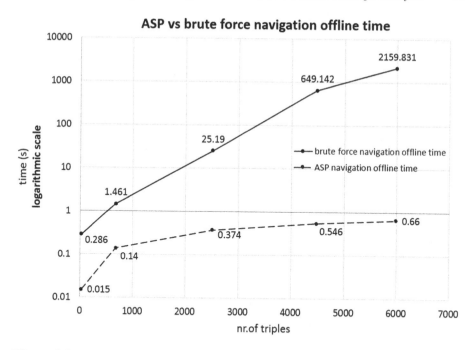

Fig. 5. Offline time for ASP vs brute force navigation tool with respect to the number of triples in the relation

tool on the logarithmically scaled y-axis in relation to the number of triples represented on the x-axis. As the chart shows, the offline time for the brute force navigation has a massive growth compared to the size of the triadic relation, while the offline time for the ASP navigation tool has a more linear growth. When comparing the average step time, the brute force navigation tool has slightly better results than the ASP navigation tool, but, as shown in Fig. 6, for subsets with less than 6000 triples the average step time is under 1 s for both approaches. Furthermore, from the experiments run on the larger data subset, containing 8133 triples, it followed that the ASP navigation tool is still usable, with an average step time of 1.194 s, as opposed to the brute force navigation tool, which turned out to have a very time consuming preprocessing phase: the Trias algorithm did not manage to compute the triconcept set in two hours.

The experiments lead us to believe that for larger datasets, the ASP navigation tool should be the preferred one, since it has a small execution time for loading the data, as well as for each navigation step, both of which are important for an interactive tool. Furthermore, in case of dynamic datasets that change frequently, it makes sense to use the ASP navigation tool which requires no preprocessing of the data.

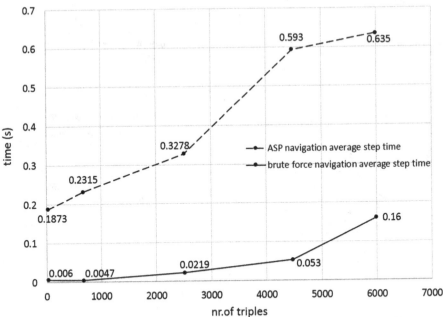

Fig. 6. Average step time for ASP vs brute force navigation tool with respect to the number of triples in the relation

7   Tool Extension

Having practical applications of FCA as a main motivation, we developed an extended tool, FCA Tools Bundle (for more details please refer to [15]), containing, besides basic FCA operations, the navigation method described in this paper as well as a different navigation method based on dyadic projections described in a previous paper [20]. From the different implementation approaches described in Sect. 5 we chose to include only the ASP based implementation in the new tool, since this method was evaluated as being more useful for large datasets.

The FCA Tools Bundle[7] currently implements features for dyadic and triadic formal concept analysis. The dyadic part contains features for lattice building and lattice visualization, while the triadic part focuses on navigation paradigms for tricontexts. The main purpose of the tool is to enable the user to visualize datasets of different dimensions as formal contexts and to interact with them. For this purpose, FCA Tools Bundle integrates different visualization formats offered by the formal contexts: concept list, concept lattice, local projections, particular formal concepts.

---

[7] https://fca-tools-bundle.com.

The graphical interface is similar to the one in the previous ASP navigation tool. For example, the same intermediate state represented in Fig. 2, can be seen in Fig. 7. In a similar manner, some of the elements are selected as being included or excluded, while the inclusions/exclusions of others ("Alouni", "HICSS", "2014") remains to be determined in the next steps of the navigation. This fact is highlighted using the message box which assists the user during the navigation: "A concept was not yet pin-pointed but the constraints have been updated to prevent reaching an invalid state.".

Fig. 7. Screenshot `FCA Tools Bundle`: intermediate state

In the final step of the navigation, the message shown to the user is updated accordingly: "A concept was found using the provided constraints.". This can be observed in Fig. 8, which is the equivalent of Fig. 3 from the previous navigation tool. Moreover, the formal concept that was reached is represented separately on its three components: extent, intent and modus.

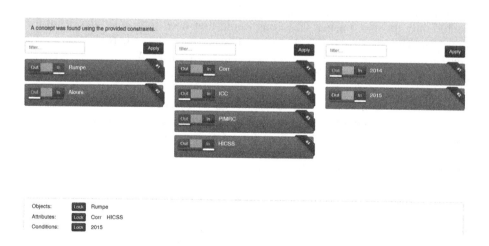

Fig. 8. Screenshot `FCA Tools Bundle`: final state

`FCA Tools Bundle` usefully integrates different navigation methods proposed for triadic formal contexts. Therefore, one can use the ASP-based navigation in order to pin-point a single formal concept and then continue using the navigation based on dyadic projections for further exploration of the tricontext. This can be seen in Fig. 8, where before each component of the formal concept, there is a button "lock". This gives the user the possibility to project the context on that particular component and look at the corresponding dyadic concept lattice. Therefore, users can easily combine navigation methods to extensively explore a large dataset.

Another very important advantage of `FCA Tools Bundle` is the usability and accessibility of the tool. While one needs prior ASP and FCA knowledge in order to use the ASP navigation tool described in Sect. 5, there is no need for prior knowledge whatsoever in order to use `FCA Tools Bundle`. The user can simply import datasets in several formats and then explore them using the FCA- and ASP-enhanced methods.

8   Conclusion

This paper presents a navigation paradigm for polyadic FCA using different implementation strategies. For higher-adic FCA, this is, to the best of our knowledge, the first navigation tool which allows to explore, search, recognize, identify, analyze, and investigate polyadic concept sets by using membership constraints, in line with the Conceptual Knowledge Processing paradigm. Experiments on 3-dimensional datasets strongly suggest that the ASP navigation approach with optimized propagation is, in general, the better choice since it has low execution times, even for larger contexts. Furthermore, in case one needs to adapt the navigation tool to an n-dimensional context for $n \geq 5$, the ASP approach is easier generalized, by following the example of the already implemented cases $n \in \{2, 3, 4\}$, whereas for the brute force navigation approach, which was implemented only for $n = 3$ using Trias, one would first need to find an algorithm for computing the n-concepts. Moreover, we have presented a new tool, `FCA tools Bundle`, which allows users who are not familiar with FCA or ASP to analyze multi-dimensional datasets and browse through maximal clusters of data.

For future work, we intend to compare the ASP approach of the n-adic case with the naive approach that uses tools such as Data-Peeler [3] or Fenster [2] which claim to be able to compute closed patterns for n-adic datasets. Also, some new features or new implementations for existing features will be integrated in the `FCA tools Bundle`. Furthermore, are planning to further investigate exploration strategies and rule mining in polyadic datasets.

Acknowlededments. Diana Troancă's research on this topic was supported by a scholarship from DAAD, the German Academic Exchange Service. The authors also thank the anonymous reviewers of an earlier version of this paper. We thank Lukas Schweizer for technical advice regarding ASP. The `FCA Tools Bundle` was implemented by Lorand Kis.

References

1. Calimeri, F., Gebser, M., Maratea, M., Ricca, F.: Design and results of the fifth answer set programming competition. Artif. Intell. **231**, 151–181 (2016)
2. Cerf, L., Besson, J., Nguyen, K., Boulicaut, J.: Closed and noise-tolerant patterns in n-ary relations. Data Min. Knowl. Discov. **26**(3), 574–619 (2013)
3. Cerf, L., Besson, J., Robardet, C., Boulicaut, J.F.: Closed patterns meet n-ary relations. ACM Trans. Knowl. Discov. Data **3**(1), 3:1–3:36 (2009)
4. Dragoş, S., Haliţă, D., Săcărea, C.: Behavioral pattern mining in web based educational systems. In: Rozic, N., Begusic, D., Saric, M., Solic, P. (eds.) Proceedings of the 23rd International Conference on Software, Telecommunications and Computer Networks (SoftCOM 2015), Split, Croatia, pp. 215–219. IEEE (2015)
5. Dragoş, S., Haliţă, D., Săcărea, C., Troancă, D.: Applying triadic FCA in studying web usage behaviors. In: Buchmann, R., Kifor, C.V., Yu, J. (eds.) KSEM 2014. LNCS (LNAI), vol. 8793, pp. 73–80. Springer, Cham (2014). https://doi.org/10.1007/978-3-319-12096-6_7
6. Ganter, B., Wille, R.: Formal Concept Analysis - Mathematical Foundations. Springer, Heidelberg (1999). https://doi.org/10.1007/978-3-642-59830-2
7. Gebser, M., Kaminski, R., Kaufmann, B., Schaub, T.: Answer Set Solving in Practice. Synthesis Lectures on Artificial Intelligence and Machine Learning. Morgan and Claypool Publishers, San Rafael (2012)
8. Gebser, M., Kaufmann, B., Kaminski, R., Ostrowski, M., Schaub, T., Schneider, M.T.: Potassco: The potsdam answer set solving collection. AI Commun. **24**(2), 107–124 (2011)
9. Gelfond, M., Lifschitz, V.: The stable model semantics for logic programming. In: Kowalski, R.A., Bowen, K.A. (eds.) Logic Programming, Proceedings of the Fifth International Conference and Symposium, Seattle, Washington (2 Volumes), pp. 1070–1080. MIT Press (1988)
10. Gelfond, M., Lifschitz, V.: The Stable Model Semantics For Logic Programming. In: Kowalski, R., Bowen, K.A. (eds.) Proceedings of the Joint International Logic Programming Conference and Symposium, JICSLP 1988, Manchester, England, pp. 1070–1080. MIT Press (1988)
11. Gelfond, M., Lifschitz, V.: Classical negation in logic programs and disjunctive databases. New Gener. Comput. **9**(3/4), 365–386 (1991)
12. Ignatov, D.I., Gnatyshak, D.V., Kuznetsov, S.O., Mirkin, B.G.: Triadic formal concept analysis and triclustering: searching for optimal patterns. Mach. Learn. **101**(1–3), 271–302 (2015)
13. Jäschke, R., Hotho, A., Schmitz, C., Ganter, B., Stumme, G.: TRIAS - an algorithm for mining iceberg tri-lattices. In: Clifton, C.W., Zhong, N., Liu, J., Wah, B.W., Wu, X. (eds.) Proceedings of the 6th IEEE International Conference on Data Mining (ICDM 2006), Hong Kong, China, pp. 907–911. IEEE Computer Society Press (2006)
14. Jäschke, R., Hotho, A., Schmitz, C., Ganter, B., Stumme, G.: Discovering shared conceptualizations in folksonomies. J. Web Semant. **6**(1), 38–53 (2008)
15. Kis, L.L., Sacarea, C., Troanca, D.: FCA tools bundle - A tool that enables dyadic and triadic conceptual navigation. In: Kuznetsov, S.O., Napoli, A., Rudolph, S. (eds.) Proceedings of the 5th International Workshop "What can FCA do for Artificial Intelligence?" co-located with the European Conference on Artificial Intelligence, FCA4AI@ECAI 2016, The Hague, The Netherlands. CEUR Workshop Proceedings, vol. 1703, pp. 42–50. CEUR-WS.org (2016)

16. Lehmann, F., Wille, R.: A triadic approach to formal concept analysis. In: Ellis, G., Levinson, R., Rich, W., Sowa, J.F. (eds.) ICCS-ConceptStruct 1995. LNCS, vol. 954, pp. 32–43. Springer, Heidelberg (1995). https://doi.org/10.1007/3-540-60161-9_27

17. Marek, V.W., Truszczyński, M.: Stable models and an alternative logic programming paradigm. In: Marek, V.M., Truszczyński, M., Warren, D.S. (eds.) The Logic Programming Paradigm: A 25-Year Perspective, pp. 375–398. Springer, Berlin, Heidelberg (1999). https://doi.org/10.1007/978-3-642-60085-2_17

18. Niemelä, I.: Logic programs with stable model semantics as a constraint programming paradigm. Ann. Math. Artif. Intell. 25(3–4), 241–273 (1999)

19. Rudolph, S., Săcărea, C., Troancă, D.: Membership constraints in formal concept analysis. In: Yang, Q., Wooldridge, M. (eds.) Proceedings of the Twenty-Fourth International Joint Conference on Artificial Intelligence (IJCAI 2015), Buenos Aires, Argentina, pp. 3186–3192. AAAI Press (2015)

20. Rudolph, S., Săcărea, C., Troancă, D.: Towards a navigation paradigm for triadic concepts. In: Baixeries, J., Sacarea, C., Ojeda-Aciego, M. (eds.) ICFCA 2015. LNCS (LNAI), vol. 9113, pp. 252–267. Springer, Cham (2015). https://doi.org/10.1007/978-3-319-19545-2_16

21. Săcărea, C.: Investigating oncological databases using conceptual landscapes. In: Hernandez, N., Jäschke, R., Croitoru, M. (eds.) ICCS 2014. LNCS (LNAI), vol. 8577, pp. 299–304. Springer, Cham (2014). https://doi.org/10.1007/978-3-319-08389-6_26

22. Voutsadakis, G.: Polyadic concept analysis. Order - A J. Theor. Ordered Sets Appl. 19(3), 295–304 (2002)

23. Wille, R.: The basic theorem of triadic concept analysis. Order - A J. Theor. Ordered Sets Appl. 12(2), 149–158 (1995)

24. Wille, R.: Methods of conceptual knowledge processing. In: Missaoui, R., Schmidt, J. (eds.) ICFCA 2006. LNCS (LNAI), vol. 3874, pp. 1–29. Springer, Heidelberg (2006). https://doi.org/10.1007/11671404_1

Highlighting Trend-Setters in Educational Platforms by Means of Formal Concept Analysis and Answer Set Programming

Sanda Dragoş, Diana Şotropa, and Diana Troancă[(✉)]

Babeş-Bolyai University, Cluj-Napoca, Romania
{sanda,dianat}@cs.ubbcluj.ro, diana.halita@ubbcluj.ro

Abstract. Web-based educational systems offer unique opportunities to study how students learn and based on the analysis of the users' behavior, to develop methods to improve the e-learning system. These opportunities are explored, in the current paper, by blending web usage mining techniques with polyadic formal concept analysis and answer set programming. In this research, we consider the problem of investigating browsing behavior by analyzing users' behavioral patterns on a locally developed e-learning platform, called PULSE. Therefore, we investigate users' behavior by using similarity measures on various sequences of accessed pages in a tetradic and a pentadic setting. and present an approach for detecting repetitive behavioral patterns in order to determine trend-setters and followers. Furthermore, we prove the effectiveness of combining conceptual scale building with temporal concept analysis in order to investigate life-tracks relative to specific behaviors discovered in online educational platforms.

1   Introduction

Nowadays, the educational system consists of two parts: the traditional educational system and the online educational system. Lately, online educational systems show a rapid development, mainly due to the growth of the Internet [21]. Analyzing web educational content is extremely important in order to help the educational process.

Online educational systems consist of techniques and methods which provide access to educational programs for students, who are separated by time and space from traditional lectures. These web-based education systems can record the students' activity in web logs, that provide a raw trace of the learners' navigation on the site [21].

It has been proven that web analytics are not precise enough for the educational content [19], as they were designed to be used on e-commerce sites, which have very different structures and requirements. However, web usage mining [26]

Diana Şotropa was supported by Tora Trading Services private scholarship.

E. Mercier-Laurent and D. Boulanger (Eds.): AI4KM 2016, IFIP AICT 518, pp. 71–92, 2018.
https://doi.org/10.1007/978-3-319-92928-6_5

provides important feedback for website optimization, web personalization [23] and behavior predictions [20].

From the teaching perspective, the online component becomes a natural extension of traditional learning. Therefore, J. Liebowitz and M. Frank define blended learning as a hybrid of traditional and online learning [18]. There are a variety of blended learning classes in universities. I.-H. Jo et al. compare on one hand the case of the discussion-based blended learning course, which involves active learner's participation in online forums, and, on the other, the case of the lecture-based blended learning course, which involves submitting tasks or downloading materials as main online activities [15]. In their paper, they show that the data collected in the first case can be analyzed in order to predict linear relations between online activities and student performance, i.e. the total score that they obtain. However, in the second case, the same analysis model was not appropriate for prediction.

It has been shown that finding a single algorithm that has the best classification and accuracy for all cases is not possible, even if highly complicated and advanced data-mining techniques are used [20]. Thus, offline information such as classroom attendance, punctuality, participation, attention and predisposition were suggested to increase the efficiency of such algorithms.

In our current research, we use formal concept analysis as a technique to discover patterns in the data logs of the educational portal. Formal concept analysis (FCA) is a mathematical theory based on lattices, that is suitable for applications in data analysis [28]. Due to the strength of its knowledge discovery capabilities and the subsequent efficient algorithms, FCA seems to be particularly suitable for analyzing educational sites. For instance, L. Cerulo and D. Distante research the topic of improving discussion forums using FCA [4,5], while our own previous contributions are focused on applying the same technique in order to analyze the user/student behavior [6–8].

This paper emphasizes how formal concept analysis tools along with answer set programming can be used for detecting repetitive browsing habits. The purpose of this research is to determine the following characteristics in the data:

- *trend-setters*, i.e., users which firstly adhere to a specific behavior and then generate a bundle of users following them;
- *followers*, i.e., users who copy the behavior of a trend-setter;
- *patterns* revealed by the occurrences of particular behaviors.

However, in order to determine trend-setters, we need to look at the data from a different perspective than the ones researched in our previous work. With that purpose, we analyze our data from a 4-adic and 5-adic perspective.

Then, we apply methods of Temporal Concept Analysis (TCA) in order to investigate in more detail the behavior of trend-setters and followers throughout the semester. TCA is also used to compare web usage patterns with respect to temporal development and occurrence.

2   Formal Concept Analysis

In this section, we briefly present the basic notions of formal concept analysis. The fundamental structures are a *formal context*, i.e. a data set that contains elements and a relation between them, and *formal concepts*, i.e. clusters of data from the defined context.

2.1   Dyadic Formal Concept Analysis and Its Extensions

FCA was introduced by R. Wille and B. Ganter in the dyadic setting, in the form of objects related to attributes [12]. In subsequent work, F. Lehmann and R. Wille extended it to a triadic setting, adding the third dimension represented by conditions [17].

Definition 1. *Dyadic formal context*
A dyadic formal context $\mathbb{K} = (G, M, I)$ is defined as a triple consisting of two sets and a binary relation $I \subset G \times M$ between the two sets. G represents the set of objects, M the set of attributes and I is called the incidence relation. The notation for an element of the incidence relation is gIm or $(g, m) \in I$ and it is read "object g has attribute m".

In order to define the notion of a formal concept, the derivation operators have to be introduced first.

Definition 2. *The derivation operator*
We define the derivation operator for the object set G and the attribute set M by: $A' = \{m \in M \mid gIm, \forall g \in A\}$ for $A \subseteq G$, and $B' = \{g \in G \mid gIm, \forall m \in B\}$ for $B \subseteq M$. For an element $g \in G$ or $m \in M$, instead of writing $\{g\}'$ and $\{m\}'$, often the notations g' and m' are used.

Based on these derivation operators, the notion of formal concept is introduced.

Definition 3. *Formal concept*
If (G, M, I) is a formal context, then a (dyadic) is defined as a pair (A, B), with $A \subseteq G$, $B \subseteq M$, $A' = B$ and $B' = A$. A is called extent and B intent of the concept. The set of all concepts of the context (G, M, I) is denoted by $B(G, M, I)$.

Definition 4. *Concept lattice*
Let (G, M, I) be a formal context and $(A1, B1)$, $(A2, B2)$ two concepts of this context. $(A1, B1)$ is a sub concept of $(A2, B2)$ if $A1 \subseteq A2$. In this case, $(A2, B2)$ is called a super concept of $(A1, B1)$. The notation is $(A1, B1) \leq (A2, B2)$. The set of all concepts with this order relation, $(B(G, M, I), \leq)$, is a complete lattice, called the concept lattice of context (G, M, I).

Definition 5. *Object concept and attribute concept*
Let (G, M, I) be a formal context, $g \in G$ an object and $m \in M$ an attribute. Then, the formal concept (g'', g') is called an object concept and it is denoted by $\gamma(g)$, while the formal concept (m', m'') is called an attribute concept and it is denoted by $\mu(m)$.

Later, G. Voutsadakis further generalized the dyadic and triadic cases to n-adic data sets, introducing the term *Polyadic Concept Analysis* [27]. Formally, an n-adic formal context is defined as follows.

Definition 6. *Polyadic formal context*
Let $n \geq 2$ be a natural number. An n-context is an $(n+1)$-tuple $\mathbb{K} := (K_1, K_2, \ldots, K_n, Y)$, where $K_1, K_2, \ldots, K_n$ are data sets and Y is an n-ary relation $Y \subseteq K_1 \times K_2 \times \cdots \times K_n$.

Formal concepts are defined as maximal clusters of n-sets, where every element is interrelated with all the others.

Definition 7. *Polyadic formal concept*
The n-concepts of an n-context $(K_1, \ldots, K_n, Y)$ are exactly the n-tuples $(A_1, \ldots, A_n)$ that satisfy $A_1 \times \cdots \times A_n \subseteq Y$ and which are maximal with respect to component-wise set inclusion. $(A_1, \ldots, A_n)$ is called a proper n-concept if $A_1, \ldots, A_n$ are all non-empty.

w3	A	B	C	D
$LT - LA$	×			
$LT - LE$	×	×	×	

w4	A	B	C	D
$LT - LA$	×			×
$LT - LE$	×	×		

Fig. 1. Visit behavior: user, chain of pages, timestamp

Example 1. Finite dyadic contexts are usually represented as cross-tables, rows being labeled with object names, columns with attribute names. Intuitively, a cross in the table on the row labeled g and the column labeled m, means that object g has attribute m.

In the triadic case, there is a ternary relation that relates objects to attributes and conditions. Here, the corresponding triadic context can be thought of as a 3D cuboid, the ternary relation being marked by filled cells. Therefore, triadic contexts can be unfolded into a series of dyadic "slices". In the following example, we consider a triadic context (K_1, K_2, K_3, Y) where the object set K_1 consists of users, the attribute set K_2 contains chains of visited pages while the conditions K_3 are the weeks of the semester when the chain occurred as a user's navigational pattern. For this small selection we obtain a $2 \times 4 \times 2$ triadic context, the "slices" being labeled by condition names.

There are exactly six triconcepts of this context, i.e., maximal 3D cuboids full of incidences:

- $(\{LT - LA, LT - LE\}, \{A\}, \{w3, w4\})$,

- $(\{LT - LE\}, \{A, B, C\}, \{w3\})$,
- $(\{LT - LE\}, \{A, B\}, \{w3, w4\})$,
- $(\{LT - LA\}, \{A, D\}, \{w4\})$,
- $(\emptyset, \{A, B, C, D\}, \{w3, w4\})$ and
- $(\{LT - LA, LT - LE\}, \{A, B, C, D\}, \emptyset)$.

The first four of these triconcepts are proper.

2.2 Many-Valued Contexts

We will briefly recall some definitions introduced by Wille in [28] regarding many-valued contexts and conceptual scaling.

Definition 8. *Many-valued contexts*
A many-valued context (G, M, W, I) consists of sets G, M, and W and a ternary relation I between G, M and W (i.e., $I \subseteq G \times M \times W$) for which it holds that $(g, m, w) \in I$ and $(g, m, v) \in I$ always implies $w = v$.

The triple $(g, m, w) \in I$ is read as "the attribute m has the value w for the object g". The many-valued attributes can be regarded as partial maps from G in W. Therefore, it seems reasonable to write $m(g) = w$ instead of $(g, m, w) \in I$.

In order to derive the conceptual structure of a many-valued context, we need to scale every many-valued attribute. This process is called *conceptual scaling* and it is always driven by the semantics of the attribute values.

Definition 9. *Conceptual scales*
*A scale for the attribute m of a many-valued context is a formal context $S_m :=$ (G_m, M_m, I_m) with $m(G) \subseteq G_m$. The objects of a scale are called **scale values**, the attributes are called **scale attributes**.*

Every context can be used as a scale. Formally there is no difference between a scale and a context. However, we will use the term "scale" only for contexts which have a clear conceptual structure and which bear meaning.

The set of scales can then be used to navigate within the conceptual structure of the many-valued context (and the subsequent scaled context). Some scales are predefined (like nominally, ordinally, etc.), while for more complex views, we need to define particular scales.

2.3 Temporal Concept Analysis

Conceptual time systems have been introduced by Wolff in [29] in order to investigate conceptual structures of data enhanced with a time layer. Basically, conceptual time systems are many-valued contexts, comprising a time part and an event part, which are subject of conceptual scaling, unveiling the temporal development of the analyzed data, object trajectories and life tracks. We briefly recall some basic definitions.

Definition 10. *Conceptual Time System*
Let G be an arbitrary set, (G, M, W, I_T) and (G, E, V, I_E) many-valued contexts. Let $\{S_m \mid m \in M\}$ be a set of scales for (G, M, W, I_T), and $\{S_e \mid e \in E\}$ a set of scales for (G, E, V, I_E). We denote by $T := ((G, M, W, I_T), (S_m \mid m \in M))$ and $C := ((G, E, V, I_E), (S_e \mid e \in E))$ the correspondent scaled many-valued contexts (on the same object set G). The pair (T, C) is called a conceptual time system on G. T is called the time part and C the event part of (T, C).

Definition 11. *Conceptual Time Systems with a Time Relation*
Let (T, C) be a conceptual time system on G and $R \in G \times G$. The triple (T, C, R) is called a conceptual time system on G with a time relation.

Definition 12. *Transitions in Conceptual Time Systems with a Time Relation*
Let (T, C, R) be a conceptual time system on G with a time relation. Then any pair $(g, h) \in R$ is called an R-transition on G. The element g is called the start and h the end of (g, h).

Definition 13. *Conceptual Time Systems with Actual Objects and Time Relation*
Let P be a set of objects, G a set of points in time and $\Pi \subseteq P \times G$ a set of actual objects. Let (T, C) be a conceptual time system on Π and $R \subseteq \Pi \times \Pi$. Then the tuple (P, G, Π, T, C, R) is called a conceptual time system on $\Pi \subseteq P \times G$ with actual objects and a time relation R (shortly a CTSOT).

For each object $p \in P$ we can be define the set $p^{\Pi} = \{g \in G \mid (p, g) \in \Pi\}$. Then the set $R_P = \{(g, h) \mid ((p, g), (p, h)) \in R\}$ is called the set of R transitions of p and the relational structure (p^{Π}, R_P) is called the time structure of p.

Definition 14. *Life track of an object*
Let (P, G, Π, T, C, R) be a CTSOT and $p \in P$. Then, for any mapping $f \colon \{p\} \times p^{\Pi} \to X$, the set $f = \{((p, g), f(p, g)) \mid g \in p^{\Pi}\}$ is called the f-life track of p.

3   Answer Set Programming for FCA

In the current paper we use answer set programming (ASP) as a method of computing formal concepts in contexts of different dimensions. ASP is a logic programming language that uses a declarative approach to solve problems [13].

In 2015, we proposed an ASP encoding that could be used to compute formal concepts and, if necessary, also add some additional constraints to the concepts [24]. We briefly describe the intuition behind this encoding and highlight the fact that it is easily extended to n-adic contexts.

Let $\mathbb{K} = (K_1, \ldots, K_n, Y)$ be an n-context. The first step encoded in the ASP program resembles "guessing" a formal concept candidate $(A_1, \ldots, A_n)$, by indicating for each element of the context if it is included in the concept or not. The second step encodes the elimination of all the previously generated candidates, for which at least one tuple is not included in the relation, i.e. $A_1 \times \ldots \times A_n \nsubseteq Y$. In the next steps, candidates that violate the maximality condition or have one empty component, need to be eliminated, ensuring that all the candidates remaining are proper formal concepts of $\mathbb{K}$. Finally, in the last step, the subset of concepts is selected, for which additional given constraints hold.

It follows from the description of the ASP encoding that it can be extended to compute formal concepts for any dimension n. After encoding the problem for a particular n-adic case, we used Clingo from the Potassco collection [14] (since it is currently the most prominent solver leading the latest competitions [3]) for running the ASP program, mainly for the grounding and solving of the encoded problem.

Furthermore, in our previous work, we presented a tool, called ASP navigation tool[1], that allows navigation through the concept space of dyadic, triadic and tetradic data sets based on the previously described ASP encoding [25]. This tool is based on membership constraints, which are encoded in the last step of the problem. For navigating with this tool, one has to choose elements of the data set and select whether they should be included or excluded from the concept. By adding such constraints, the tool ensures that, eventually, one will get to a final state of the navigation, which corresponds to a proper formal concept, i.e. a real data cluster. The implementation of the ASP navigation tool is described in more details in our previous work [24,25].

In our current analysis we extend the ASP encoding to pentadic data sets and compute formal concepts in order to analyze correlations between tetradic clusters of data in a 5-adic setting and hence obtain interesting patterns, such as trend-setters. Moreover, after analyzing pentadic patterns, we use the previously mentioned tool to take a closer look at some of the students that stand out in the obtained results.

4   Web Usage Mining on PULSE

Educational environments can store a huge amount of potential data from multiple sources, with different formats, and with different granularity levels (from coarse to fine grain), or multiple levels of meaningful hierarchy (keystroke level, answer level, session level, student level, classroom level, and school level) [22]. Therefore, an important research direction focuses on developing computational theories and tools to assist humans in extracting useful information from the rapidly growing volumes of data [21]. In Web Mining, data can be collected server side or client side, through proxy servers or web servers logs.

The logged data needs to be transformed in the data pre-processing phase into a suitable format, on which particular mining techniques can be used. Data-preprocessing contains the following tasks: data cleaning, user identification, session identification, data transformation and enrichment, data integration and data reduction [21]. Data cleaning is one of the major pre-processing tasks,

---

[1] https://sourceforge.net/projects/asp-concept-navigation.

through which irrelevant log entries are removed, such as crawler activity. For the next steps of the pre-processing phase, more data transformations are necessary, such as data discretization and feature selection, in order to perform user and session identification, data integration from different sources and to further analyze the data.

The usage/access data considered for this analysis is collected from the web logs of an e-learning portal called PULSE [10]. PULSE records the entire activity of its users and, although it has more types of users, we are currently interested only in the students' activities. PULSE also records other individual information about students such as the academic results, or users' attendances to the laboratories, which in our university system is mandatory.

We will briefly present the entities which are representative for our study:

- The **user** is a student that accesses web files through a browser. Users can be uniquely identified by their login ID (educational content on PULSE can be accessed only after a login phase);
- A **session** is an actual HTTP session;
- A **chain** is defined as a chronologically ordered sequence of visited pages during a session;
- The **timestamp** is the date and time of the access.

The relationship between different entities can be determined by temporal aspects hidden in the data. Data may describe developments over time or temporal mechanisms (i.e. time series data) or it may reveal the patterns that evolve over time [16]. Finding evolving patterns is an important challenge which plays a key role in the process of understanding users behavior.

In order to determine users behavioral patterns we consider chains of pages (sequences of visited pages where the accessed page becomes the referrer for the next one). These chains are formed on the assumption that the temporal order of clicks describes the path the user takes through the web site. When the referrer is not the same as the last page accessed it means that the user opened a new browser tab or window, and we called that part of our session chain a new branch.

We have compared chains of the same user in order to determine each student's repetitive behavior. We also compared chains for different users to identify the influence one user may have over another and to get relevant information in order to determine possible trend-setters. For comparing these chains, we have used the Jaccard, Cosine and Sorensen similarity measures [11].

Among these measures, the Cosine Similarity has the advantage of a smaller complexity. Let A and B be two chains, we build the occurrence vectors C_A and C_B for each chain . This similarity measure returns the cosine of the angle between the two occurrence vectors, by the following formula:

$$CosineSimilarity = \frac{C_A \cdot C_B}{||C_A|| \cdot ||C_B||}, \tag{1}$$

where $||C_A||$ is the magnitude of the vector C_A.

For all these tests, the order in which access classes occurred into a chain has not been taken into consideration. Therefore, having P_1; P_2 two pages which were visited by a user in the same session, we consider that "page P_1 is visited just before page P_2" has the same weight as "the page P_2 is visited just before page P_1" while we compute chain similarity.

Given that the Cosine Similarity algorithm has the smallest complexity, requires less memory space and it is computational more efficient than the other similarity measures mentioned, we decided to use this method in order to find similar patterns of usage behavior within the same time sequence or among different time sequences. Next we considered only the chains that have a Cosine similarity of at least 80%.

For each student we determine chains of pages visited during a visit/session and associate them to the corresponding week based on the visit's timestamp. The next step of the analysis is to compare the chains of a user amongst each other, in order to determine students' repetitive behavior. Furthermore, we compare chains of different users in order to identify the influence one user may have over another, and to get relevant information for identifying possible *trend-setters*, as defined in our previous work [6].

For the experiments presented in this paper, we consider a group of students from the same program, studying the same subject. For this group, containing 23 students, we logged every single file access of every student, for a specific subject, over a period of one academic semester.

The pages of the e-learning platform are grouped by their content into classes. Our interests for this analysis focuses on 10 of these classes which contain pages related to the educational content. These classes are described and denoted in Table 1.

Table 1. Classes of pages we are interested in

#	Code	Description
1	I	**Information about** the lecture and the laboratories
2	PE	**Information about** the practical examination
3	WE	**Information about** the written examination
4	L	Overview information about all the **lectures**
5	Ls	Slides and notes for all **lectures**
6	LP	Overview information about the test papers given during **lectures**
7	LPs	Details on all **lecture** test papers
8	LA	**Lab** assignments
9	LE	**Laboratory** examples
10	LT	Theoretical support for all **laboratories**

The first three classes presented in the Table 1 contain general information related to the way the lecture and laboratories are conducted and about the examination procedure. The next four classes are related to Lectures, while the last three classes are related to the Laboratory activity.

Using the Cosine similarity measure with a threshold of 80%, we obtain pairs of students having similar behavior, i.e. similar chains in a proportion of at least 80%. That behavior occurs for each student in a certain week. Thus, we have pairs of two students, a common behavior, and the corresponding weeks in which the behavior occurred for each student. Therefore, for each student X, we can construct a tetradic context, containing all students that have similar behavior with X as objects, the actual behaviors as attributes and weeks as conditions and states. Herefrom, for a student X, a tetradic concept (A_1, A_2, A_3, A_4) can be understood as follows: all students in A_1 have, in comparison to student X, similar behaviors to the ones described by the chains in A_2; however, this behaviors occur in the weeks $w_1 \in A_3$ for student X, while for the students in A_1 they occur in the weeks $w_2 \in A_4$.

In order to reduce the granularity of our behavior, which, at this point, is a chain, we substitute all chains with binary codes denoting the presence in that chain of the 10 access classes that we are interested in. Moreover, to reduce redundancies, we consider an additional constraint, mainly that the timestamp when the behavior occurs for student X should be previous (or identical) to the timestamps when it occurs for the other students. We eliminate the tuples for which the constraint does not hold, hence making sure that a common behavior appears in the context corresponding to a single user, mainly the user initiating that behavior, and not the students that "learn" that behavior.

We will refer to this student as *trend-setter* and to the others having the behavior in common as *followers*. In this context setting, we are able to determine bundles of users that have similar behavior. A similar detailed analysis of such user bundles, based on different techniques, was published in a previous paper [7].

5   Identifying Trend-Setters Based on Navigational Patterns

In the current paper we would like to analyze our data from the 5-adic perspective, in order to determine trend-setters, i.e. students that create a behavior that is assimilated by others, influencing them in the way they use PULSE. Our approach is to extend the tetradic concept by aggregating all previously described 4-adic contexts into one 5-adic context. Therefore, we introduce a new dimension, called *state2*, that corresponds to the set of users. In the pentadic context, *state* becomes *state1*, in order to avoid any confusions. For computing the pentadic concepts of the described context, we use the ASP program described in Sect. 3.

The timestamp constraint mentioned above determines that all concepts will contain followers as objects and trend-setters as state2, for a specific behavior in the attribute set. The condition set contains the occurrence weeks of that behavior for the trend-setter, while the state1 set contains the occurrence weeks of that behavior for the followers.

For this set of experiments[2], we consider for each trend-setter the behaviors containing the maximum number of classes, from the 10 classes we are interested in. We will denote this behavior as *rich behavior*.

In the obtained results, we observe several patterns. In order to analyze each pattern separately, we have grouped the clusters by behavior/pattern (attribute) and sorted them by the week when it first occurred (condition). Thus, we represent each pattern in a different table and observe the corresponding trendsetters. The first such pattern is presented in Table 2. Here, student F can be identified as a trend-setter, if, for a particular behavior (e.g. "Ls-LP-LT-LA"), he/she is the first to have that rich behavior, and the other students have an 80% similar behavior[3] in the same or the following weeks.

Table 2. Sample of 5-adic concepts grouped by behavior, trend-setter and week when it occurred

Objects	Attributes	Conditions (w1)	States 1 (w2)	States 2
F, B, D, H, Q	Ls-LP-LT-LA	4	4	F
D	Ls-LP-LT-LA	4	4, 6, 7	F
...	Ls-LP-LT-LA	4	...	F
X	Ls-LP-LT-LA	4	8	F
S	Ls-LP-LT-LA	4	13	F
...	Ls-LP-LT-LA	4	...	F
G	Ls-LP-LT-LA	8	15	X
S	Ls-LP-LT-LA	13	13	S
L,W	Ls-LP-LT-LA	13	14	S
G	Ls-LP-LT-LA	13	15	S

Using these criterias for grouping the data, we mainly obtain groups with different behaviors and their corresponding trend-setters. These trend-setters are often different from each other and initiate the behavior in different weeks. However, there is one particular case that stands out and that can be observed in the subset of data represented in Table 2. Here, we can see that the behavior "Ls-LP-LT-LA" has three potential trend-setters, mainly students F, X and S. The behavior occurs for student F in week 4, for student X in week 8 and for student S in week 13. Although they can all be seen as trend-setters for a particular group of students, we deduce that the real trend-setter is student F, since he is the first to have this behavior. Moreover, the other two students are considered to be his followers, as it can also be seen in the 4th and 5th line of Table 2.

Another aspect that is notable for all the results is that most behaviors are focused on the *Lecture* and *Laboratory* access classes and these behaviors

---

[2] More details about the analyzed data and the obtained results can be found at http://www.cs.ubbcluj.ro/~fca/tests-for-ai4km/.

[3] The similarity was performed on actual session chains as defined in Sect. 4.

Table 3. Followers that become trend-setters for enhanced behaviors

Objects	Attributes	Conditions (w1)	States 1 (w2)	States 2
F	WE-PE-LT-LA	5	5	F
E	WE-PE-LT-LA	5	6	F
M	WE-PE-LT-LA	5	9	F
D	WE-PE-LT-LA	5	12	F
Q	WE-PE-LT-LA	5	13	F
U,W	WE-PE-LT-LA	5	14	F
V	WE-PE-LT-LA	5	17	F
Q	LPs-WE-PE-LT-LA	13	13	Q
U,W	LPs-WE-PE-LT-LA	13	14	Q
V	LPs-WE-PE-LT-LA	13	17	Q
U,W	LPs-WE-PE-LT-LA	14	14	U

Table 4. Behavior initiated by two trend-setters

Objects	Attributes	Conditions (w1)	States 1 (w2)	States 2
B, Q, D	I-WE-PE-Ls-LPs	16	16	B, Q
D	I-WE-PE-Ls-LPs	16	16, 17	B, Q
D, V	I-WE-PE-Ls-LPs	16	17	B, Q
O, Y	I-WE-PE-Ls-LPs	16	20	B, Q

are initiated either at the beginning of the semester, i.e. in weeks 3 and 4, or towards the end of the semester, i.e. in weeks 10 and 12.

The second pattern, that we observed, represents behaviors which are assimilated by other students, who then enrich this behavior and become themselves trend-setters for the new enhanced behavior. This pattern is depicted in Table 3. Here, we can see that user F is the trend-setter for behavior "WE-PE-LT-LA". User Q learns this behavior, adds a new access class to it, "LPs", and becomes trend-setter for the new behavior. Then, we can observe that user U learns the new behavior and becomes a follower of user Q.

Another interesting aspect that can be observed in Table 4 is, that there can be two trend-setters initiating the same behavior. Here, we can observe that students B and Q are both trend-setters for the same behavior "I-WE-PE-Ls-LPs".

Next, we focus on behaviors that have no followers. We call such behaviors *singular*. It turns out that these behaviors contain longer chains of distinct classes than behaviors that have followers, being complex, repetitive behaviors of some students. The subset of concepts corresponding to these behaviors is represented in Table 5. The results show that these behaviors reoccur only in the same week of their initiation and for the same user. Hence, there are no actual followers for those rich behaviors. This can also be deduced from the fact that the object sets of the concepts contain only the user that initiated that behavior, while for other behaviors, that have followers, these can be seen in the object set (see Table 2). The longest chain observed here contains 8 classes out of the 10 that we are interested in, and the average length of the chains in the behaviors from Table 5 is 7.

Table 5. Longest chains of classes that occur in student's behavior

Objects	Attributes	Conditions (w1)	States 1 (w2)	States 2
H	L-Ls-LP-WE-LT-PE-LA-I	4	4	H
C	L-Ls-LP-LT-LE-LA	5	5	C
C	LPs-WE-LT-LE-PE-I	11	11	C
D	Ls-LPs-WE-LT-LE-PE-LA-I	12	12	D
I	L-Ls-LP-LPs-WE-LT-PE	14	14	I
M	L-Ls-LPs-WE-LT-PE-LA	9	9	M

In what follows, we present some statistics regarding the number of followers that each trend-setter has, the number of weeks in which the behavior occurs and the size of the chain in the corresponding behavior, in terms of number of "interesting" classes. These statistics are represented in Fig. 2. Here, for each distinct behavior, i.e., chain, we represented different entries of some users. Therefore, student A has two distinct behaviors on which he/she has followers. We denoted these instances with A1 and A2. Similar, we have two instances for B, three instances for F, and two instances for Q.

As it can be seen in Fig. 2, student F had the most followers (16 different students) for a particular behavior containing 4 important classes, behavior that reoccurs in 11 weeks. Furthermore, we observe that there seems to be a directly proportional relation between the number of followers and the number of weeks in which the behavior reoccurs for each trend-setter. However, there also seems to be an indirect proportional relationship between the number of followers and the size of the chain in a behavior.

In what follows, we focus on user F, which did not stand out in any of the different analyses that we have run on the same data, but using different techniques (presented in our previous work [6,7]). However, this user stands out in the current analysis, for example by having the largest number of followers. The surprise is even greater, since this student attended only 5 out of 14 laboratories and his/her academic results are below average. In order to further analyze the behavior of this particular user, we return to the tetradic approach and use the concept navigation tool based on ASP[4]. Therefore, we continue our investigation on the entire data set (i.e., not only the similar entries) in order to determine more details about the behavior of user F. Thus, we navigate through concepts

---

[4] https://sourceforge.net/projects/asp-concept-navigation.

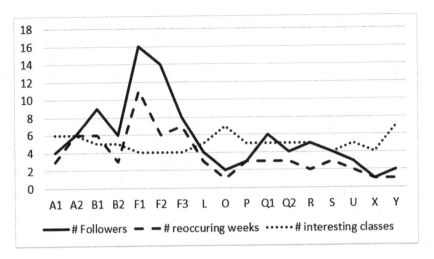

Fig. 2. Statistics regarding behaviors and followers

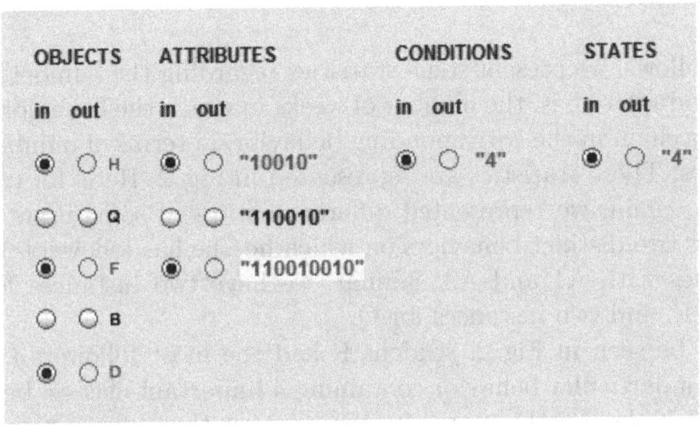

Fig. 3. Generated cluster for behavior "Ls-LP-LT-LA" and user F

corresponding to the rich behaviors previously observed for user F. As a first example, we start by choosing the behavior "Ls-LP-LT-LA", which for the data analysis is encoded as "110010010". Next, we choose the user F as an object, meaning that we are looking for all the weeks when user F repeated this behavior and what other users or behaviors belong to this data cluster. As shown in Fig. 3, this behavior occurs only in week 4, but also for users H and D. Moreover, the group of users H, F and D have another behavior in common that occurs in the same week, mainly "LT-LA", i.e. "10010". We can see in Fig. 3, that although the state represented is an intermediate state, we can already discover patterns in the data. The fact that it is an intermediate state is determined by the objects which are neither "in", nor "out" of the data cluster, meaning we did not reach a formal concept yet because the maximality condition is not satisfied.

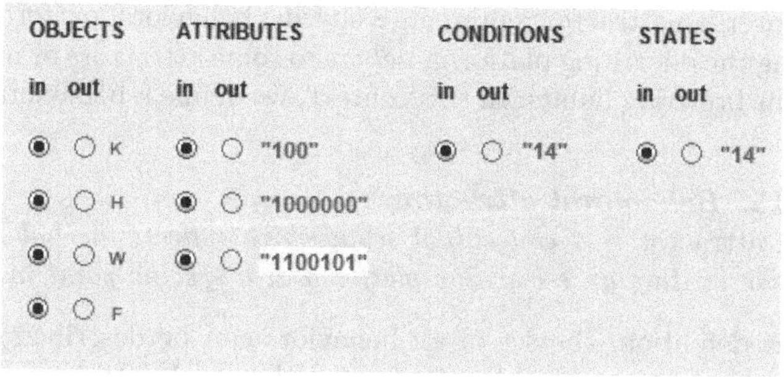

OBJECTS		ATTRIBUTES			CONDITIONS			STATES		
in	out	in	out		in	out		in	out	
◉	○ K	◉	○	"100"	◉	○	"14"	◉	○	"14"
◉	○ H	◉	○	"1000000"						
◉	○ W	◉	○	"1100101"						
◉	○ F									

Fig. 4. Generated cluster for behavior "LPs-WE-PE-I" and user F

For the second example, as depicted in Fig. 4, we choose a different rich behavior for F, mainly "LPs-WE-PE-I", which is encoded as "1100101", and again student F as an object. This behavior turns out to occur in week 14 and it is a common behavior for users K, H and W. Furthermore, this group of users also has in common the behaviors "PE", i.e. "100", and "LPs", i.e. "1000000". Here we have reached a formal concept, as all objects are included in the cluster and there are no more inclusions/exclusions to be determined.

Concluding our analysis, we state that trend-setters and followers of particular behaviors can be identified in a pentadic setting as described earlier in this section. However, in order to take a closer look at the behavior of certain users, it is useful to go back to a tetradic setting and explore correlations of their behaviors and the weeks in which the behaviors occur for the same or for other users. Using the visual navigation tool, one can further explore the data and find potential new patterns which were not revealed by the pentadic context that we have analyzed. Furthermore, the ASP navigation tool can be extended to n-adic datasets, in order to visualize patterns in pentadic or higher-adic contexts.

6   Temporal Aspects: Dynamics and Relationships

Graphically represented conceptual hierarchies prove to be a very efficient tool for the discovery and understanding of complex relationships between knowledge units. ToscanaJ [1,2] offers the possibility of using available diagrams according to one's interest. Therefore, one can use diagrams aggregation in order to investigate the existence of patterns in attributes correlation. Different scenarios can be formed using only a small subset of the diagrams (the same diagram can even be included more than once).

In another paper ([9]) we have presented an investigation of user behavior in educational platforms using Temporal Concept Analysis, where *attractors* were introduced and defined as sets of scales in conceptual time systems. As the name suggests, an attractor is either influencing or describing the users behavior in the educational platform. Therefore, attractors prove to be special categories of scales which need to be related to specific time granules, when the attractor occurs. Moreover, an attractor represents a specific behavioral pattern. Students, while browsing the e-learning platform, *adhere* to some attractors or not, showing thus particular browsing habits. In this context, we define a behavioral attractor as follows.

Definition 15. *Behavioral attractor*
A behavioral attractor *is a conceptual scale which reflects the habits of a student/user while visiting an e-learning platform at a specific point in time.*

Given this definition, the set of all behaviors can be described by a set of conceptual scales on time granules. So, each behavioral pattern represents the event part of the conceptual time system at the specific time granule.

We build users' life tracks by setting time granularity at week level and marking the temporal trajectory of the student through rich behaviors. Then, we consider for every time granule one of the rich behaviors presented in Table 2 and 3 (i.e. "Ls-LP-LT-LA", "WE-PE-LT-LA", and "LPs-WE-PE-LT-LA") and build users' life tracks superimposing them on the concept lattice of the corresponding *behavioral attractor*.

In order to emphasize users' life tracks we need to set up some criteria and an order. Then we need to load the appropriate scale in a user defined order. For instance, if we select the following scales in this order: the login scale (a nominal scale which contains user IDs), the weeks scale (a nominal scale which contains the academic weeks), and the corresponding *behavioral attractor*, we obtain a complete view of the trend-setters' and followers' activity for the three patterns: pattern "Ls-LP-LT-LA" is shown in Fig. 5, pattern "WE-PE-LT-LA" in Fig. 6, and pattern "LPs-WE-PE-LT-LA" in Fig. 7.

Behavioral attractors are *unintended attractors*, which according to our previous research ([9]) are crystallizing behavioral patterns showing how users are using the resources, independently to the intention of the educator. *Behavioral attractor* is an umbrella concept that also encompasses different types of behaviors, some of them defined in our previous paper ([9]), such as *habitual attractor* (i.e., the habit of branching - opening more browser windows or tabs during a visit) and *critical attractor* (i.e., many page accesses that last only a few seconds in critical time intervals such as examination). In this paper we will refer to the navigational (unintended) behavior as to what students are more likely to access during a visit.

In what follows we define a subclass of behavioral attractors called *rich attractors* which expresses the behaviors containing the maximum number of classes.

Definition 16. *Rich attractors*
A rich attractor *is a conceptual scale on a given time granule which reflects the behavior containing the maximum number of classes out of the 10 classes we are interested in (LA, LT, LE, L, Ls, LP, LPs, WE, PE, I). This attractor reveals clues about some other unintended patterns showed off by the users in order to collect educational information.*

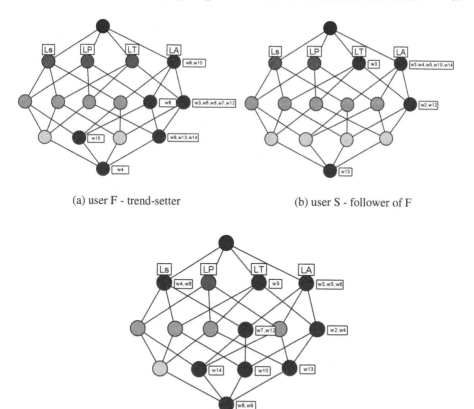

(a) user F - trend-setter   (b) user S - follower of F

(c) user X - follower of F

Fig. 5. Users' life tracks according to the sample of 5-adic concepts grouped by behavior and trend-setter as presented in Table 2

Further on we use the above described formalizations in order to see in more detail the evolution of the students during the entire semester relative to specific behaviors. Figure 5 presents the first such example. We start with the *rich attractor*, that is the scale corresponding to the behavior ("Ls-LP-LT-LA") generated by the *trend-setter* F. On this scale we build life tracks superimposing the behavior on different weeks on the concept latice of the rich attractor. Figure 5 contains the life tracks of the *trend-setter* F and two of his *followers*: X and S.

We deduce that the results obtained using TCA and presented in Fig. 5 are similar to the ones obtained with the help of Polyadic FCA and presented in Table 2. Figure 5(a) presents the life track of the trend-setter, i.e. how user F is using the educational content considered for this particular rich behavior ("Ls-LP-LT-LA") on the entire semester (i.e., 14 weeks). Analysing the corresponding conceptual scales for users S and X (see Figs. 5(b) and (c)) which, as we have seen, are followers of user F, gives us insights about the relation between trend-setters and followers. In Fig. 5(b) we observe that, after follower S assimilates the trend-setter's behavior in week 13, in week 14 he goes back to his usual behavior and visits only pages from the "LA" class. Follower X, however, adheres to the rich attractor in week 8, and then continues to visit pages from all the classes of the rich behavior "Ls-LP-LT-LA". This indicates that the new behavior of X was, in this case, influenced by this *rich attractor.*

As depicted in Fig. 5(b), there are multiple stages in which a user can be placed over time. There are cases in which users are starting by visiting pages from a single class. That is the case of student S which starts visiting one of the classes, i.e. *LT*, included in the considered rich behavior in week 3. This habit is quite normal for a first encounter with the course content. Still, there are situations in which, given a certain point in time, the fact that a student visited pages from a single class might signalize a superficial approach to the learning process, which needs to be corrected by the instructor. However, the more deeper a life track goes in the concept lattice, the more related content is visited, and thus, it might be assumed that more specific skills related to the subject are acquired and the overview on the learning topic is better. This reflects how seriously users are approaching a specific subject, its structure being unveiled by the corresponding rich attractor. These facts motivate also our interest for rich behaviors.

The *trend-setter* F adheres to the *rich attractor* in the 4th week of the semester. This is represented in the lattice from Fig. 5(b) in the lowest node. In the same figure it can be observed that the user F has a tendency to focus more, over time, on lab assignment (LA) pages. He/she maintains over the weeks a more comprehensive behavior than the *followers* (i.e., most of his/her behavior is found on the lower nodes), focusing however on laboratory-related material, and less on the lecture-related classes. The *followers* have a different behavior over the time than the *trend-setter*. However, they seem to have a very similar behavior among themselves. Moreover, they seem to pay more attention to the lecture-related material.

Figure 6 presents the life track of user F as a trend-setter of another rich behavior ("WE-PE-LT-LA") and the life track of user Q, i.e., one of his followers for the corresponding behavior. Moreover, Fig. 7 depicts the newly generated rich attractor in which Q becomes the trend-setter of an enhanced behavior ("LPs-WE-PE-LT-LA") and other followers are identified. By looking at Fig. 6 one might say that the habits of the follower seem, at a first glance, not to be influenced by the learned rich behavior. However, by analyzing both figures we can observe how followers learn the behavior of the trend-setter, and then, deviate from the trend-setter's behavior by introducing new rich behaviors.

As depicted in Fig. 6(a), *trend-setter* F is interested in all the four classes (i.e., WE, PE, LT, LA) throughout the semester. On the other hand, Fig. 6(b) shows that *follower* Q seems to have a "one time" rich behavior in week 13, than returning to his old patterns of accessing only pages related to Laboratories (i.e., LT and LA). However, if we project Q's behavior on the new attractor (as depicted in Fig. 7(a)),

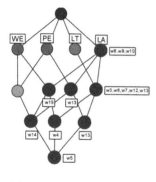
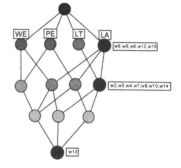

(a) user F - trend-setter   (b) user Q - follower of F

Fig. 6. Users' life tracks according to the sample of 5-adic concepts grouped by behavior and trend-setter as presented in Table 3

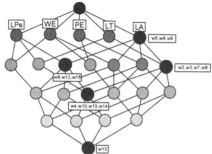
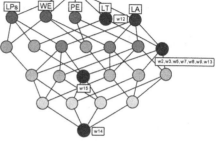

(a) user Q (follower of F) extends F's behaviorand becomes trend-setter for the new behavior.

(b) user U learns the new behavior and becomes a follower of user Q

Fig. 7. Users' life tracks of followers that become trend-setters for enhanced behaviors as presented in Table 3

it can be observed that Q's behavior contains in the latter weeks pages form the Lecture Papers (LPs) class and not only pages related to Laboratories.

User U, follower of Q on the extended behavior "LPs-WE-PE-LT-LA" (see Fig. 7(b)) has again a less comprehensive behavior as Q, apart from weeks 14 and 15.

If we project the behavior of F on the extended Q's behavior, as depicted in Fig. 8, we see that although he/she has visited pages in LPs class, it has no visit containing pages from all 5 classes. However, F has a comprehensive behavior as his behavioral patterns are represented on the lower nodes of the latice.

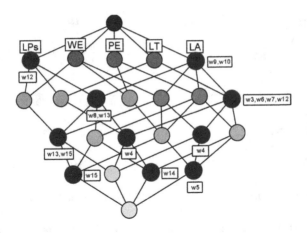

Fig. 8. user F; the initial trend-setter on the extended behavior

7   Conclusions and Future Research

Web is an excellent tool to deliver educational content in the context of an online educational system, while web mining is an efficient technique that can be used to find valuable information in the data. While statistical analysis, through its quantitative approach, might give insight information about web traffic, we believe that formal concept analysis, through its qualitative approach, reveals the potential of hidden patterns inside web logs. Our research is focused on discovering useful patterns that lead to a more efficient interaction between the users and the platform, and that help students acquire the necessary knowledge during the learning process more easily. In this paper, we propose a new method for investigating trend-setters based on pattern extraction from Web log files. We have analyzed students that initiate a behavior that is eventually assimilated by other students, influencing them in the way they use the portal. This analysis helps educators understand the users' behavior and use the obtained knowledge for optimizing and personalizing the e-learning portal. We have also investigated how new navigational patterns initiated by the trend-setters influence the behavior of the followers in time. Moreover, we analyzed the evolution of a bundle of users, over time by applying temporal concept analysis on the data set corresponding to the users that showed up in previous tests. Life tracks give valuable feedback to the instructor regarding how the online educational resources are used over time. They also might be helpful for analyzing the usability of the online educational content, and eventually for improving the structure of the platform and developing new educational instruments. We intend to continue this research, considering pattern structures, relational FCA and other FCA varieties.

References

1. Becker, P., Correia, J.H.: The ToscanaJ suite for implementing conceptual information systems. In: Ganter, B., Stumme, G., Wille, R. (eds.) Formal Concept Analysis. LNCS (LNAI), vol. 3626, pp. 324–348. Springer, Heidelberg (2005). https://doi.org/10.1007/11528784_17

2. Becker, P., Hereth, J., Stumme, G.: ToscanaJ: An open source tool for qualitative data analysis. In: Duquenne, V., Ganter, B., Liquiere, M., Nguifo, E.M., Stumme, G. (eds.) Proceedings of the Workshop Advances in Formal Concept Analysis for Knowledge Discovery in Databases, FCAKDD 2002, co-located with the European Conference on Artificial Intelligence, ECAI 2002, Lyon, France, pp. 1–2 (2002)

3. Calimeri, F., Gebser, M., Maratea, M., Ricca, F.: Design and results of the fifth answer set programming competition. Artif. Intell. **231**, 151–181 (2016)

4. Cerulo, L., Distante, D.: Topic-driven semi-automatic reorganization of online discussion forums: a case study in an e-learning context. In: Global Engineering Education Conference (EDUCON), pp. 303–310. IEEE (2013)

5. Distante, D., Fernandez, A., Cerulo, L., Visaggio, A.: Enhancing online discussion forums with topic-driven content search and assisted posting. In: Fred, A., Dietz, J.L.G., Aveiro, D., Liu, K., Filipe, J. (eds.) IC3K 2014. CCIS, vol. 553, pp. 161–180. Springer, Cham (2015). https://doi.org/10.1007/978-3-319-25840-9_11

6. Dragoş, S., Haliţă, D., Săcărea, C.: Attractors in web based educational systems a conceptual knowledge processing grounded approach. In: Zhang, S., Wirsing, M., Zhang, Z. (eds.) KSEM 2015. LNCS (LNAI), vol. 9403, pp. 190–195. Springer, Cham (2015). https://doi.org/10.1007/978-3-319-25159-2_18

7. Dragoş, S.-M., Haliţă, D.-F., Săcărea, C.: Distilling conceptual structures from weblog data using polyadic FCA. In: Haemmerlé, O., Stapleton, G., Faron Zucker, C. (eds.) ICCS 2016. LNCS (LNAI), vol. 9717, pp. 151–159. Springer, Cham (2016). https://doi.org/10.1007/978-3-319-40985-6_12

8. Dragoş, S., Haliţă, D., Săcărea, C., Troancă, D.: Applying triadic FCA in studying web usage behaviors. In: Buchmann, R., Kifor, C.V., Yu, J. (eds.) KSEM 2014. LNCS (LNAI), vol. 8793, pp. 73–80. Springer, Cham (2014). https://doi.org/10.1007/978-3-319-12096-6_7

9. Dragoş, S.-M., Săcărea, C., Şotropa, D.-F.: An investigation of user behavior in educational platforms using temporal concept analysis. In: Bertet, K., Borchmann, D., Cellier, P., Ferré, S. (eds.) ICFCA 2017. LNCS (LNAI), vol. 10308, pp. 122–137. Springer, Cham (2017). https://doi.org/10.1007/978-3-319-59271-8_8

10. Dragos, S.: PULSE - a PHP utility used in laboratories for student evaluation. In: International Conference on Informatics Education Europe II (IEEII), Thessaloniki, Greece, pp. 306–314 (2007)

11. Gan, G., Ma, C., Wu, J.: Data clustering: theory, algorithms, and applications, vol. 20. SIAM, Philadelphia (2007)

12. Ganter, B., Wille, R.: Formal Concept Analysis. Springer, Heidelberg (1999). https://doi.org/10.1007/978-3-642-59830-2

13. Gebser, M., Kaminski, R., Kaufmann, B., Schaub, T.: Answer Set Solving in Practice. Synthesis Lectures on Artificial Intelligence and Machine Learning. Morgan and Claypool Publishers, San Rafael (2012)

14. Gebser, M., Kaufmann, B., Kaminski, R., Ostrowski, M., Schaub, T., Schneider, M.T.: Potassco: the potsdam answer set solving collection. AI Commun. **24**(2), 107–124 (2011)

15. Jo, I.H., Park, Y., Kim, J., Song, J.: Analysis of online behavior and prediction of learning performance in blended learning environments. Educ. Technol. Int. **15**(2), 71–88 (2014)

16. Kriegel, H.P., Borgwardt, K.M., Kröger, P., Pryakhin, A., Schubert, M., Zimek, A.: Future trends in data mining. Data Min. Knowl. Disc. **15**(1), 87–97 (2007)

17. Lehmann, F., Wille, R.: A triadic approach to formal concept analysis. In: Ellis, G., Levinson, R., Rich, W., Sowa, J.F. (eds.) ICCS-ConceptStruct 1995. LNCS, vol. 954, pp. 32–43. Springer, Heidelberg (1995). https://doi.org/10.1007/3-540-60161-9_27

18. Liebowitz, J., Frank, M.: Knowledge management and e-learning. CRC Press, Boca Raton (2010)

19. Macfadyen, L.P., Dawson, S.: Numbers are not enough. why e-learning analytics failed to inform an institutional strategic plan. Educ. Technol. Soc. **15**(3), 149–163 (2012)

20. Romero, C., Espejo, P.G., Zafra, A., Romero, J.R., Ventura, S.: Web usage mining for predicting final marks of students that use moodle courses. Comput. Appl. Eng. Educ. **21**(1), 135–146 (2013)

21. Romero, C., Ventura, S.: Educational data mining: a survey from 1995 to 2005. Expert Syst. Appl. **33**(1), 135–146 (2007)

22. Romero, C., Ventura, S.: Wiley Interdisciplinary Reviews: Data Mining and Knowledge Discovery. Data Min. Educ. **3**(1), 12–27 (2013)

23. Romero, C., Ventura, S., Zafra, A., de Bra, P.: Applying web usage mining for personalizing hyperlinks in web-based adaptive educational systems. Comput. Educ. **53**(3), 828–840 (2009)

24. Rudolph, S., Săcărea, C., Troancă, D.: Membership constraints in formal concept analysis. In: Yang, Q., Wooldridge, M. (eds.) Proceedings of the Twenty-Fourth International Joint Conference on Artificial Intelligence, IJCAI 2015, Buenos Aires, Argentina, July 25–31, 2015, AAAI Press, pp. 3186–3192 (2015)

25. Rudolph, S., Săcărea, C., Troancă, D.: Conceptual navigation for polyadic formal concept analysis. In: Proceedings of the 4th International Workshop on Artificial Intelligence for Knowledge Management AI4KM at IJCAI, pp. 35–41 (2016)

26. Spiliopoulou, M., Faulstich, L.C.: WUM: a tool for web utilization analysis. In: Atzeni, P., Mendelzon, A., Mecca, G. (eds.) WebDB 1998. LNCS, vol. 1590, pp. 184–203. Springer, Heidelberg (1999). https://doi.org/10.1007/10704656_12

27. Voutsadakis, G.: Polyadic concept analysis. Order **19**(3), 295–304 (2002)

28. Wille R.: Conceptual landscapes of knowledge: a pragmatic paradigm for knowledge processing. In: Gaul W., Locarek-Junge H. (eds.) Classification in the Information Age. Studies in Classification, Data Analysis, and Knowledge Organization. Springer, Heidelberg (1999)

29. Wolff, K.E.: Temporal concept analysis. In: ICCS-2001 International Workshop on Concept Lattices-Based Theory, Methods and Tools for Knowledge Discovery in Databases, Stanford University, Palo Alto (CA), pp. 91–107 (2001)

Selection of Free Software Useful in Business Intelligence. Teaching Methodology Perspective

Mieczysław Owoc[✉] and Maciej Pondel

Wroclaw University of Economics, Wrocław, Poland
{mieczyslaw.owoc,maciej.pondel}@ue.wroc.pl

Abstract. Modern decision-taking processes are supported by advanced information technologies. There are many products on the market representing more and more smart solutions therefore selection of proper software is not easy especially if managers are oriented on minimizing costs in computer infrastructure. It is significantly important for people representing small and medium-sized enterprises. On the other hand it is common expectation of well-educated staff as graduates of academia. This is very essential assumption for educational sector where teaching methodology and defined software packages are discussed and proposed. Initial point of the research is discussion of teaching methodology essence and diversification. In our paper we propose methodology of selection free software tools essential in education limited to teaching of business intelligence. Especially Magic Quadrant prepared by Gartner is carefully analyzed. The main software products offering by different companies were taken into account and procedure of selection was defined including list of criteria essential in the choice of a product. The list of criteria embraces different perspectives of the selection. Power BI as the selected tool is presented in more details.

Keywords: Business Intelligence teaching · Free software
Methodology of teaching · Supporting of decision-making

1 Introduction

Business Intelligence (BI) approach and technology seems to be inherent part of current business sector. That is why it is an essential part of teaching program on the Business Informatics field of study and other relate. Main goals of learning Business Intelligence (considered in the completed project in the Erasmus+ framework DIMBI: *Developing the innovative methodology of teaching Business Informatics*) are:

- understand the objectives of Business Intelligence systems implementation,
- understand the process of knowledge acquisition from data,
- gather skills necessary to collect project stakeholders requirements in terms of BI implementation,
- understand the most important features and capabilities of BI tools,
- be able to map the requirements to the capabilities of BI tools,

E. Mercier-Laurent and D. Boulanger (Eds.): AI4KM 2016, IFIP AICT 518, pp. 93–105, 2018.
https://doi.org/10.1007/978-3-319-92928-6_6

- gather skills essential to solve most common problems and issues concerning implementation of Business Intelligence,
- practise implementation of Business Intelligence solution in defined business perspectives and selected IT tools.

Authors assume that detailed recognition of one selected IT tool should not be considered as the objective of BI course. If a student understands the overall process and is familiar with a specific tool he or she is able to utilise those skills quickly learning other systems that he or she meets in practise. Of course it would be beneficial if a student uses the same tool in a real projects that he or she learns during studies but we do not intend to teach students all possible IT tools that are available on the market. During the research on various BI systems we found that the essential aspects of all systems are relatively similar in terms of their functionality and ways of usage and it is quite simple to transfer student's experience from one tool to another.

The process of IT tools selection should start from definition of criteria that fully support assumed objectives. The most important criterion is existence of free edition of the tool. Authors of the paper find it crucial that system is available for free. Even most advanced and advantageous system can be considered useless in learning process if it is not easy accessible. The aim of our research project is to deliver the course that is available in a regular form of classroom training and also as an online course. The course will be available for every willing university or students.

The licence limits can discourage potential students from using the tool and from attending the whole prepared course. That is why the availability of a free version was in our case the must. Authors understand that IT tools can be delivered in a various models and free version in each case can have a different meaning. Existence of a free version can signify that (see Laurent 2004):

- Tool is available in an open source model (GNU GPL, LGPL or Mozilla Licence)
- Tool is available under MIT, BSD, Apache or Academic Licence
- Tool is available as commercial one but it has free edition that contains defined limitations.

Regarding limitations – providers of those tools can define them on a different levels. Some limit available features of software other limit the amount of data that can be processed on a free tool. From authors' perspective limitation of system's feature is of course a meaningful disadvantage. Limitations on other levels like data capacity or others restricting availability of the tool usage in enterprise environment (e.g. integration with corporate AD, number of concurrent data sources) may be considered less important and be acceptable for learning purposes.

2   Teaching Methodology of Business Intelligence

Modern higher education faces with many problems and challenges. Basically, it is result of continuous civilizational progress and growing up student's requirements.

There are many opinions presenting challenges in this matter, for example P. Raj keeps the following: education for all, education cost and education leveraging jobs

(Quora 21). Therefore society, economics and marketing aspects are considered. On the other side higher education in globalization world according to Robertson should shaped competitive abilities of students (see Robertson 2010).

All the mentioned factors have the real impact on **teaching methodology** independently of topics learned. Teaching methodology as the defined set of methods "comprises the principles and methods used by teachers to enable student learning" (https://en.wikipedia.org/wiki/Learning) (Wikipedia/teaching_methods) is strictly connected with techniques and tools useful in development of student competencies and abilities.

It is very important especially in the courses where the usage of software tools is the must. Business intelligence is an example of such discipline.

Main objectives of teaching business intelligence can be defined as follows:

- understanding the objectives of Business Intelligence systems implementation,
- understanding the operation of knowledge acquisition from data warehouse models,
- gathering skills necessary to collect project stakeholder requirements in terms of BI implementation,
- understanding the most important characteristics and capabilities of BI tools,
- being able to map the requirements to the capabilities of BI tools,
- gathering skills essential to solve most common troubles and matters concerning implementation of Business Intelligence,
- practical implementation of Business Intelligence solutions in selected business perspectives and selected IT tools.

All of the mentioned objectives should be considered in Business Intelligence phases context. Assuming BI process as a composition of crucial phases: define, integrate, analyze and visualize BI - teaching methods and supporting tools should embrace one or more phases. General idea of the process is presented in the Fig. 1.

The nature of the phases introduced into BI forces the use of computer tools throughout the cycle. It is critical for integration of data sources, analysis of big data and visualization results in many forms.

Teaching methodology applied for obtaining BI educational objectives can be formulated as a set of traditional and innovative methods of delivering knowledge and developing student competences. There is a big number of formulated teaching methods prepared in more general applying or oriented on particular domains (see Haroun 2017) classified as teacher- or student-centered, direct instructions and many others. Postulated and checked for BI methods are depicted in Fig. 2.

The list of suggested methods is a bit disputable but expresses the most frequently applied solutions starting from traditional lectures up to more creative ways of training student capabilities like performing projects or solving problems with BI tools. Some of them (marked with "$\sqrt{}$" symbol) require advanced software tools. It is especially important in case of BYOD ("bring your own device") formula – where students are obliged to use their personalized technologies during education. Similarly "Solving problems using BI tools" method requires from students the usage of adequate software tools in order to integrate data sources, analyze them and visualize results helpful in the defined quests.

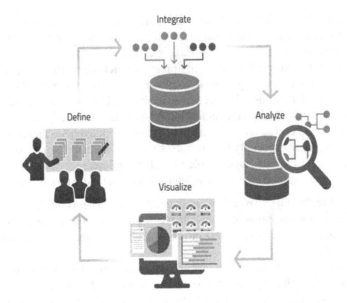

Fig. 1. Phases of BI process. *Source:* http://www.tacticaledge.us/services/software-development/
analytics-and-business-intelligence/

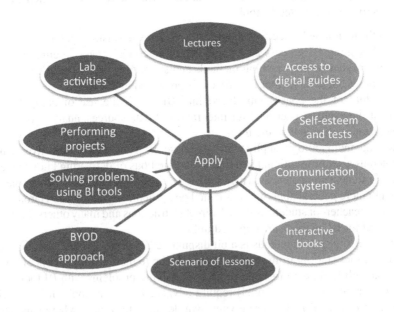

Fig. 2. Methods assumed in teaching of BI. *Source: own elaboration*

The tools accompanying teaching methodology can support more than one phase (for example analysis and visualization) and be strictly connected with the defined methodology. The main problem in selection of the tool is fulfill potential users requirements.

3   Related Works and the Selection Method

The problem of selection right tools that can be applied for particular users is very old and still very actual, especially when market of these products is very rich. Therefore managers looking for appropriate software should take into account many aspects including functionality, easiness of the usage, supporting and cost of software acquisition and maintenance. There are several approaches and guidelines that can be useful in the selection of software tools. Below is a list of examples offering relatively objective approach to select the discussed software:

- Survey on evaluation and selection the right BI tool (Sherman 2015) where
- Survey on selection BI tools (BITSu 2016) where independent group of professionals depicted about 200 criteria aggregated in 12 aspects allowing for selections the best solution for an user,
- Selection of BI tools (BITSe 2016) where comparison of selected platforms is discussed e.g. Business Objects against Oracle Hyperion or IBM Cognos,
- Presentation of BI products (BISP 2016) where characteristics of business intelligence software is delivered from quality communication and resources points of view,
- Best practices for BI tools selection (BP4BI 2006) where Authors describe very useful approach covering the whole selection process starting from the selection of business champion up to make an Informed Business decision focused on financial aspect of this decision,
- Team expert is preparing solution for customers (BI Verdict 2016), including analyst and customer opinions in the context of balanced perspective on BI tools market.

One of the most prestigious companies presenting surveys on different IT products is Gartner group – the report on BI tools covers five different cases and essential capabilities of the discussed software (GartnerBI 2016).

Main use cases refer to the following aspects:

- Agile Centralized BI Provisioning – stressing the self-contained data management capabilities of the particular platform e.g. an agile IT-enabled workflow,
- Decentralized Analytics – essential when it is capability of a creation workflow from data to self-service analytics,
- Governed Data Discovery – denotes IT-managed content including at least: governance, reusability and promotability,
- Embedded BI – a workflow embraces embedded BI content,
- Extranet Deployment – including analytical content availability for external customers.

The presented cases exhaust main possible options of usability BI tools limited to modern technological workflows.

Capabilities of the analyzed BI tools can be evaluated using several criteria. Criteria set has been defined in the following groups:

- Infrastructure - covering four criteria; BI platform administration, Cloud BI option, security and user administration and connectivity to structured as well as unstructured data sources.
- Data Management – taking into account different activities important from administration point of view expressed as three criteria: governance and metadata management, self-contained ETL and data storage processes,
- Analysis and Content Creation – a list of key (from analytical point of view) four criteria including: embedded advanced analytics, analytic dashboards, interactive visual exploration and mobile exploration and authoring,
- Sharing of Findings – referring to three output criteria: embedded analytic content, publishing analytic content and collaboration and social BI, essentially improving presentation of the analytical results.

The presented fourteen criteria apart of the mentioned five cases were used to demonstrate capabilities of important on BI market products traditionally visualized as four quadrants representing niche players, visionaries, challengers and leaders (see Fig. 3).

Fig. 3. Magic quadrant for business intelligence and analytics platforms. *Source:* (GartnerBI 2016).

In the Figure all sorts of producers are present commercial as well as so-called free-software. Actually just three platforms are located in the quadrants Leaders – Tableau,

Qlik and Microsoft; there are some versions of the products available as free of charge packages. Microsoft Power BI is representative of software that cover all necessary functionalities as BI and Analytics platform. Note also, that some packages offers capabilities essential in the ETL process – so they may work as Data Warehouse software (for example SAS, SAP and Pentaho products).

Underwood in his research explores 2017 Gartner BI Magic Quadrant Results stressing changing roles of some producers and monitoring tendency of leader's proposal (Underwood 2017).

Interesting characteristics of tools limited to open-source category is presented by Lindsay Wise (see Wise 2012). In this essay these platforms are termed as Open-Source Business Intelligence (OSBI) and contain four crucial components: Data Warehouse, Data Integration, Analytics Engine, and Reporting combined with Dashboard functionality. According to OS and BI solutions particular platforms should fulfil business as well as technological requirements; convergence approach in supporting such project is a must.

Keeping in mind all the mentioned approaches we have to focus on teaching Business Intelligence aspects in the selection of appropriate platforms. Two dimensions in the elaborated method must be defined: list of free of charge BI Analytic Tools and criteria of evaluation the particular platforms. So, we've defined free of charge platforms as the subcategory of discussed earlier non-commercial and open-source platforms.

Our research was limited to the following platforms:

- SAP Lumira (http://saplumira.com/)
- Qlik Sense (http://www.qlik.com/products/qlik-sense)
- Tableau (http://www.tableau.com/)
- MS Power BI (https://powerbi.microsoft.com/en-us/)
- Pentaho (http://www.pentaho.com/)
- Jaspersoft BI (http://www.jaspersoft.com/)

We've tested the itemized platforms through performing BI projects by students in the Data Warehouse (DW) and Business Intelligence courses or as topics of their Bachelor or Master Thesis.

The second dimension – the criteria set – represents selected capabilities essential from educational point of view. Taking into account earlier presented we've focused on the following aspects: general look, intuitiveness, easiness of implementation, safety, support, BI analysis, data sources, other functionality, and two educational perspectives: DW and BI.

4 Results of Comparative Analysis

The presented in the previous section selection method has been applied to evaluate of usability the included tools in teaching Business Intelligence topics. Teams of students familiar with particular platforms prepared reports on fulfil software capabilities using defined criteria. They voted independently under supervision of tutors. In some cases they needed additional explanation about particular criteria. For example general look was mostly identified with the interface offered by producers. At least two teams were appointed to grade one platforms. We used scale 0 (the worst) to 5 (the best) grades for evaluation purpose. Most of the criterion defined acquired qualitative measures (general look, easiness of implementation) while the others quantitative (additional functionality or licence). Comparative analysis of the evaluated platforms is presented in Table 1. Apart of the defined earlier criteria two additional capabilities were included: educational aspects separately for teaching Business Intelligence and Data Warehouse.

Table 1. Evaluation of BI analysis tools

BI analysis tools free of charge						
Criterion	SAP Lumira	Qlik Sense	Tableau	MS Power BI	Pentaho	Jaspersoft BI
Look	4	4	4	5	3	4
Intuitiveness	4	4	5	4	3	3
Easiness of implementation	3	4	4	4	2	3
Safety	2	5	3	5	3	2
Support	2	4	4	4	0	3
BI Analysis	5	5	3	5	4	3
Data Sources	1	4	3	5	5	4
Additional Functionality	4	3	3	5	5	3
Licence	2	5	3	4	5	2
Educational aspect (DW)	3	4	3	3	5	5
Educational aspect (BI)	4	4	3	5	4	3
Summary	34	46	38	49	39	35

Source: own elaboration

The highest ranked was MS Power BI as the platform with 6 best scores (especially important considering BI educational aspect. Lower position of this platform in DW educational point of view can be compensated through the usage of the second platform. In our opinion combination of two tools: MS Power BI and Pentaho looks very promising.

In the next Figs. (4 and 5) the platforms are presented stressing educational capabilities (for BI and DW respectively) in the context of summary evaluation.

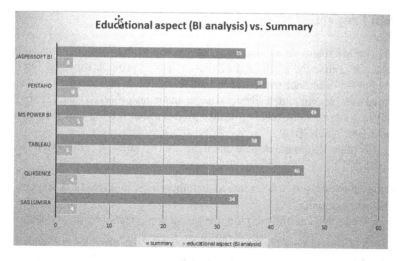

Fig. 4. Educational aspect of BI analysis vs. Summary. Source: own elaboration

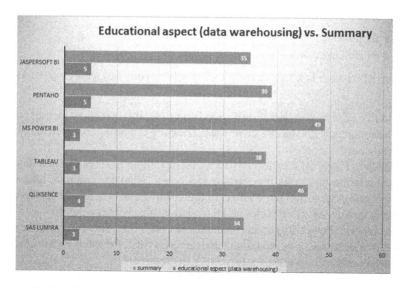

Fig. 5. Educational aspect of DW vs. Summary. Source: own elaboration.

The best position belongs to **MS Power BI**, which was elected as a leader for BI education. Additional characteristics of this tool is presented in the next section.

The best position of the second elected platform Pentaho is confirmed on this Figure.

5   Main Properties of Power BI

Power BI is a Microsoft developed suite that belong to the group of tools called Self Service BI. Of course the main goal of such tool is to provide Business Intelligence capability but the difference expressed in Self-service boils down to the fact that in this tool the final user (accountant, analyst, manager and many other) is capable to build their own analysis (data models, reports, dashboards) without relying on the help from IT department (see Webb 2014).

Also, it is important to emphasize that there are two kinds of self-serve BI user (see Pondel 2015):

- Analytics Power Users who create visual apps from multiple data sources – both internal and external.
- Regular Users that can fully explore the visual apps created by power users or IT.

Power BI is a cloud-based business analytics service that provides user with the most important BI features like creating rich interactive reports with Power BI Desktop and monitoring the health of business using live dashboards. It includes 2 main approaches to analyse data:

- Power BI Desktop,
- Power BI for Office 365.

Power BI desktop is a free desktop tool in which you can (see Power BI Desktop 2016):

- Import. You can import data from a wide variety of data sources. After user connects to a data source, he or she can shape the data before importing to match analysis and reporting needs.
- Model data. Power BI Desktop provides data modelling features like autodetect and manual relationships definition, custom measures, calculated columns, data categorization, and sort by column. There is Relationship View, where user gets a customizable diagram view of all tables and the relationships between them.
- Create reports. Power BI Desktop includes Report View. There user can select the fields he wants to display, add filters and choose visual. Prepared visualisations are interactive and authors consider them very impressive.
- Save. Power BI Desktop, allows user to save work as a Power BI Desktop file (pbix).
- Publish. Power BI Desktop, allows user to publish and share prepared datasets and reports to Power BI site (that is a cloud based service).

Power BI for Office 365 is a cloud based service available via web browser that allows (see Power BI service 2016):

- Execution of similar report creation process like in Power BI desktop (import, model, create report).
- Connect to services. An user is able to connect to content packs for a number of services such as Salesforce, Microsoft Dynamics, and Google Analytics. Power BI uses user's credentials to connect to the service, and then creates a Power BI

dashboard and a set of Power BI reports that automatically show data and provide visual insights.

- Create Dashboards. They are personalized and provide user capability to monitor most important data, at a glance. A dashboard combines on-premises and cloud data in a single, consolidated view across the organization.
- Sharing the data. In Power BI an user can share dashboards, reports, and tiles in several different ways e.g. Publish a report to the web, share a dashboard with associates, create a dashboard in a group, then share it with co-workers outside the group.
- Q&A in Power BI. Capability of processing natural language user's question and receive answers in the form of charts and graphs.
- Quick Insights. Power BI searches different subsets of dataset while applying a set of algorithms to discover potentially-interesting insights. Power BI scans as much of a dataset as possible in an allotted amount of time. Example algorithms are: Majority, Category outliers, Overall trends in time series, Correlation and many more.

Example Power BI dashboard is presented on the Fig. 6, where solving of real problem can be illustrated in many ways.

Fig. 6. Power BI example dashboard. Source: own work *on* https://app.powerbi.com

The presented dashboard is the final result of all phases included to BI process; students can select appropriate data, perform analysis employing appointed algorithms and visualize survey outcomes using different charts.

6 Conclusions

The general findings of the research can be formulated as follows:

- Innovation of teaching methods relies on application of sophisticated software tools supporting all phase of BI process,
- There are plenty platforms that can be used in teaching of Business Intelligence. Quite promising capabilities are offered as free software thus these tools are real competitors for commercial products,
- For better selection of the appropriate platform the list of criteria should be elaborated. Existing rankings are very useful in formulation of the set of capabilities essential for individual users,
- The presented method of the selection platforms for teaching Business Intelligence can be applied in teaching courses devoted to advanced technologies supporting business,

Our further research can be devoted to prepare multidimensional knowledge base supporting the selection procedure.

References

BISP: Business Intelligence software products (2016). http://www.bitool.net

BITSe: Business Intelligence Tools Selection (2016). http://www.bi-dw.info/bidw56.html

BITSu: Business Intelligence Tools Survey (2016). http://www.businessintelligencetoolbox.com

BIV: Business Intelligence Verdict (2016). http://www.bi-verdict.com/

BP4BI: Best Practices for Business Intelligence Tool Selection (2006). http://www.b-eye-network.com/view/2578

Cupoli, P., Devlin, B., Ng, R., Petschulat, S.: ACM Tech Pack on Business Intelligence/Data Management. ACM (2013)

GartnerBI: Gartner: Magic Quadrant for Business Intelligence and Analytics Platforms (2016). https://www.gartner.com/doc/reprints?id=1-2XXET8P&ct=160204

Haroun (2017). http://www.bchmsg.yolasite.com/methods.php

Jaspersoft BI: http://www.jaspersoft.com/

St. Laurent, A.M.: Understanding Open Source and Free Software Licensing. O'Reilly Media, Inc., Sebastopol (2004)

Martinez, M.: Gartner Positions Microsoft as a leader in BI and Analytics Platforms for ten consecutive years (2017). https://powerbi.microsoft.com/en-us/blog/gartner-positions-microsoft-as-a-leader-in-bi-and-analytics-platforms-for-ten-consecutive-years/

MS Power BI (2016). https://powerbi.microsoft.com/en-us/

Passioned Group: Business Intelligence Tools Survey (2017). https://www.passionned.com/kb/business-intelligence-tools-survey/

Pentaho (2016). http://www.pentaho.com/

Pondel, M.: A concept of enterprise Big Data and BI workflow driven platform. In: 2015 Federated Conference on Computer Science and Information Systems (FedCSIS). IEEE (2015)

PowerBI Desktop (2016). https://powerbi.microsoft.com/en-us/documentation/powerbi-desktop-get-the-desktop/

Power BI service (2016). https://powerbi.microsoft.com/en-us/documentation/powerbi-service-basic-concepts/

Qlik Sense (2016). http://www.qlik.com/products/qlik-sense

Quora 21: Blog on challenges of higher education. https://www.quora.com/What-are-the-challenges-for-higher-education-in-the-21st-century

Robertson, S.L.: Challenges Facing Universities in a Globalising World (2010). https://susanleerobertson.files.wordpress.com/2012/07/2010-robertson-challenges.pdf

SAP Lumira (2016). http://saplumira.com/

Sherman, R.: How to evaluate and select the right BI tool (2015). http://searchbusinessanalytics.techtarget.com/feature/How-to-evaluate-and-select-the-right-BI-analytics-tool

Tableau (2016). http://www.tableau.com/

Underwood, J.: (2017). http://www.jenunderwood.com/2017/02/22/2017-gartner-bi-magic-quadrant-results/

Webb, C.: Power Query for Power BI and Excel. Apress (2014)

Wikipedia/teaching_method: https://en.wikipedia.org/wiki/Teaching_method

Wise, L.: Using Open Source Platforms for Business Intelligence. Avoid Pitfalls and Maximize ROI. Morgan Kaufman (2012)

Internet Platform for City Dwellers
Based on Open Source System

Łukasz Przysucha[✉]

Wroclaw University of Economics, Komandorska 118/120, 53-345 Wroclaw, Poland
lukasz.przysucha@ue.wroc.pl

Abstract. Smart City is a relatively new idea developed by many business entities, organizations and residents. The world's population migrates to cities. Experts estimate that more than 65% of the global GDP will be produced by the 600 largest cities in the world in 2025. It definitely motivates to increase the intensification of activities aimed at development of urban systems which improve the efficiency of the residents and entrepreneurs working there. The author focused on a characterization and theoretical creation of the Internet Platform which will synchronize all the processes occurring in the smart city and support the development of a new ideas related with improving life in the city. The main goal of this article is to find a common ground between smart city and CMSes based on GNU GPL license which through the properties are open to development by thousands of users, fully flexible and customized for any urban area. The author suggested 10 areas of smart city which may be reflected in the portal. Usage and integration all of them may revolutionize the look on development of cities in the world.

Keywords: Smart City · Platform · City dwellers · Content management system
Open source · GNU GPL

1 Introduction

The number of operations performed on the Internet continues to increase year-on-year. The network has a huge functionality and impact on society. Considering the application in the metropolitan area Internet can be a tool for a broad social communication, source of information similar to radio, newspapers, alarm for citizens in crisis situations and many more influencing lives easier. Recently there is much discussion about smart solutions in terms of the city but solutions in practice look pale. Often this is due to high costs of implementation of systems, lack of comprehensive tools and ignorance managing the city. The following article aims to propose examples of innovative solutions for cities that will be in the platform for city dwellers based on one of the free content management systems using the GNU GPL. The author believes that the use of solutions that are currently being standardized globally, developed by millions of users and free should be implemented to create the concept of smart.

E. Mercier-Laurent and D. Boulanger (Eds.): AI4KM 2016, IFIP AICT 518, pp. 106–118, 2018.
https://doi.org/10.1007/978-3-319-92928-6_7

2 Smart Society

The concept of smart city, depending on the various definitions and scopes can include many aspects of life. According to this basic smart city can be called the area (city, region, agglomeration), which consists of four elements [1]. The first of these are creative residents who are "enlightened" their activities and use their knowledge as well as develop it. Another pillar is effectively working organizations and institutions processing existing knowledge [2]. On the technical side must be ensured adequate technological infrastructure - broadband cable services in the network, digital space for data and remote tools for knowledge management. The final element is the ability to innovation. Komninos N. explains this as part of management and the ability to solve problems that appear for the first time since the innovation and management under uncertainty are key to assessing intelligence. Smart society may be divided into more detailed groups according to aspects of:

(1) Smart business and economy - innovative environment for entrepreneurs. It must be flexible and easy in the law. Creative start-ups, new ideas and refreshed businesses are classified for a special care. There is a high level of productivity. Relationship of the economy are both against the local as well as global. There is also a high degree of flexibility in the labor market.

(2) Human capital [3] - educated people who are not afraid to implement new ideas. A wide variety of force. The universities educate specialists in their fields. In addition to running his own family and work, people have time for social activities and are open to others. They participate actively in community life.

(3) Management - the power is handed over to the residents. Urban sector is divided between education, work and residents. Residents have easy ability to communicate.

(4) Mobility - clean and efficient transport system, an integrated system for traffic control.

(5) Environment - development planning, smart resource management, investments in renewable energy sources, clean the foreground.

(6) Quality of life - safe neighborhood, a high level of health care, easier access to public services, tourist attractions, good housing conditions and constantly revised level of prosperity (Fig. 1).

The author of this work is that the main consolidator of the above factors can be innovative platform for city dwellers. It could integrate all the parts together and facilitate the development of each individual. It is true that there are many websites dedicated for metropolitan areas but they have the assumption transmit only messages from the press and public offices. They do not allow interaction between residents. Considering the wider national IT clusters providing facilities information they have similar assumptions - to integrate the participants, strengthen innovation and technology, and to build a strong position compared to other countries. Their main objectives are the integration of the national information technology, promotion of sustainable urban development, promotion and implementation of the concept of smart grid, creating an area of cooperation between IT companies, universities, and government support for innovation in

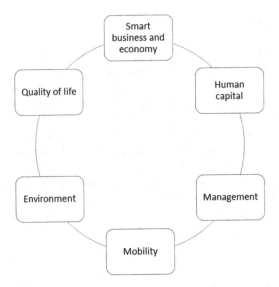

Fig. 1. One of the proposed schemes of smart city.

the economy, the acquisition of human capital and knowledge, and many others. With the full cooperation between research centers, managers of cities and private companies it is possible to fully implement the concept of intelligent city. It is important that the idea of smart city was developed around the world, and that the resources allocated to it is still growing, because, according to forecasts [4] from 2010 until 2025. 65% of the global GDP will be produced by the 600 largest cities in the world. Therefore, implementation of the idea of smart city can have a direct impact on increasing the efficiency of our civilization.

In recent years many start cities refer to ecology [5]. Clean water, air and soil related issues are discussed at numerous international conferences and reports. Please note that in order to implement ecological aspects, rules and standards, cities must keep the rules and standards and implement appropriate technologies.

3   The Concept of Internet Platform Based on Smart City

As already noted, Smart City is largely composed of several components [6]. The people of the Smart community are not the only ones, but they are also supporting their technology and institutions. Strategic directions in key dimensions are also important. There are strategic principles that, once fulfilled, can change the shape of the Smart Society to a much more advanced.

The first one is the integration of technological factors. This methodology includes not only the focus on the technical aspects that hook up the infrastructure, machinery and equipment. It is well known that hardware issues are essential to the functioning of many entities and projects, but considerable attention should be paid to the assumptions and functioning of systems. The concept of urban portal discussed in the article carries many risks. It is first dealing with the complexity of many elements. For example, downloaded data from devices that use city navigation will be in real-time mode for an interactive map that will define traffic in the streets. This is a huge complexity of the information being sent. In the case of several million cities, even if some of the inhabitants will be using the application at that time, a large amount of data will be generated and processed. Based on these reports, further traffic reports may be generated in the city and proposed further solutions to eliminate congestion and traffic jams in the streets. Another example may be a map of the actual temperature that will be transmitted immediately using a drones network that makes meteorological images. In many cases the cities are large, vary in height, and can be met with completely different measurements in one district and another across the city. Thanks to such images and accurate real temperatures, residents would be sure that the temperature and atmospheric conditions of a given location are accurate to the street. Examples may be more, but it is worth noting the data obtained from this type of project. It should be standardized in one format so that it can be aggregated and used for further work. Technology in Smart City is not only the ability to improve the standard of living of the people, but also the ability to estimate activities and create solutions for the future.

There are many solutions such Internet Platforms dedicated for urban areas. They have the usual basic features such as the provision of information from the council and news about events happening at the moment. The author considers it necessary to increase the range of functionality and make a real connection to streamline all the paths of smart city that the portal could be a tool and an integral root of the functioning of the entire smart city. It is also important that the portal could be based on an open source license. This option lets locals to develop the website the tools used in the building can be created by thousands of users and provide maximum safety.

It is worth noting that the core of the system may be available for every city the same, while extensions and functionality will be completely different. Such implementation will be very specific in comparison with other web projects. It should be noted that a large number of factors and dynamic variables are included in the platform. In this case, infrastructure and data flow will play a very important role.

In the case of use of solutions of a private company the city is dependent on a group of several people who support the project without the involvement of people from outside. More about the use of the GNU GPL and application it during the implementation in the next section.

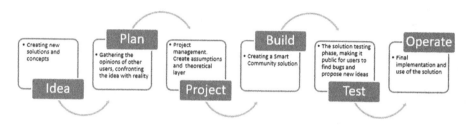

Fig. 2. Process of designing Smart solution.

The issue of implementing solutions is also whether the implementation process of the solution will be coordinated, and if so to what extent by the city authorities. Work planning and concept modification can be outsourced to outsourcing companies, but it will generate additional costs and city managers may lose control of the implementation process at some point. Below is the process of designing Smart solutions for the city (Fig. 2).

Another concept is to create a regional platform for city dwellers for example dedicated for one country. The server room and technical support team would be in one location. The "cloud" solution can have both advantages and disadvantages. Most cloud computing benefits are in terms of IT cost savings. In general the lack of on-premises infrastructure removes associated operational costs in the form of power, air conditioning and administration costs. One place in a given area will certainly minimize costs, however, the security issue will be discussed in this aspect. Although cloud service providers implement the best security standards and industry certifications, storing data and important files on external service providers always opens up the risks. If services are provided by an external company under the rules of outsourcing, principals should pay close attention to the security of the whole system, as the data on the links will sometimes contain confidential information. Cloud computing provides enhanced and simplified IT management and maintenance capabilities through central administration of resources, vendor managed infrastructure and SLA backed agreements. All updates in the IT infrastructure and maintenance area are eliminated, as all resources are maintained by the service provider.

The following Fig. 3 shows the cycle of information flow from the place of obtaining data to the server room.

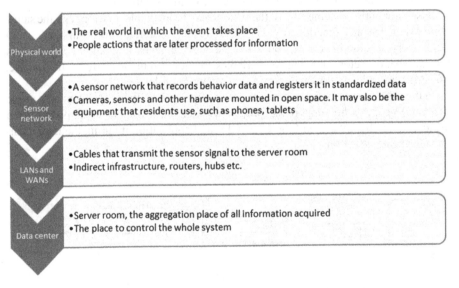

Fig. 3. Elements required to Smart City platform.

4   Content Management Systems

Content management systems are dynamic pages. At the beginning it is worth to define the concept of static and dynamic pages, and determine the differences between them. Classic pages had a heyday in the 90s of the last century. Currently abandoned standardization of static pages and dynamic websites became more popular. Static pages are those portals [7], which do not change their content when calling in your browser. In order to introduce any changes on the page, the administrator is forced to overwrite the files manually. Simple part, based on HTML have both advantages and disadvantages. They are quite simple to prepare, and the whole process of creation of the site is fast. On the market there are free programs-wizards that allow you to modify portal without knowing the language. Preparation static page does not require much effort; thus it is quite cheap. The biggest use of such sites are simple business websites. They do not require a server with PHP and MySQL databases. Hosting can also be free. Unfortunately, in the case of sending the page to the server and placing on the modification, it is necessary to have a basic knowledge of creating websites. An important drawback is the lack of interaction with users. These sites serve only to provide information unilaterally without any action on the line user-administrator. They are therefore usually less interesting than the dynamic and users spend less time on them. Dynamic pages are generated in real time in front of the server HTML based on data provided by the program for browsing the Internet. These sites are dependent on the actions taken by the user is currently browsing them. For example, after adding the comment on the page there is a new entry, date added and author. Sometimes they are also given additional user identification data such as IP, browser, which was displayed page and the version of the system. You can change the content in two ways [8] - the first method, client-side use of scripting languages such as JavaScript and ActionScript, which make direct changes to the elements (DOM. Document Object Model). The main benefits of this method are shorter response times, much less load on the server, and a better effect of the interactive application. It is not required to interact with the database, making it easier to change the code. The second method is called server-side, using programming languages such as PHP, ASP and Perl. This processing is useful for a contact database and a permanent memory. Example of the operation is to validate the user or data exchange.

A content management system is software [9] that allows you to create, manage and publish content. Early adoption of CMS contained mainly on the management of documents and files, usually in the internal forum, now is the management of content in the network at the level of the public. The purpose of such systems is to provide an intuitive interface for viewing content users, as well as the interface for the administrator of the site, usually with the position of the panel administrator. CMSes are great help in the way of work with the system [10] (Fig. 4):

(1) Dynamic content.
(2) Easy to make changes.
(3) Tab for content management.
(4) Addition of interactive content.
(5) Integration with the media.

(6) Full control over the entire site.
(7) Allowing multiple people management side.

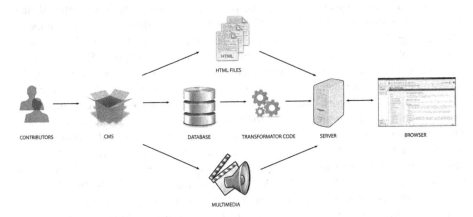

Fig. 4. Place of CMS during the page request

Content management systems offer dynamic content operations. With WYSIWYG editors such as the administrator does not need to have knowledge of HTM and other conditions necessary to deposit the content on the page. All interference in the content is done through a web browser. Readers are kept informed about the authors, dates of creation and modification of articles.

Add, update, delete, copy selected parts are extremely quick and easy. The administrator can divide the roles of action moderators and editors according to the due rights. Thanks to attributes, the person who is the editor will get the opportunity to log on only to the content of the selected departments. It has the possibility of penetration of the system functions available to administrators the entire site. In this case, the tab content management is the administration panel. The most CMS has a fairly clean interface, intuitive user interface buttons and functions, as well as the standard visual system. Systems support an interactive content such as discussions, class schedules for students, staff cards or surveys. This is thanks to support for multiple languages. For example, the CMS can be written in PHP, but the content on the site are added in HTML or script administrator deposited in JavaScript. This openness gives also a large integration with the media. On the one side can be placed graphics, video, advertising in Flash technology and many others. A super administrator with all privileges can control the entire site and all users. In addition to the admin panel is a panel important place on the server and access to ftp. New updates and documents can be sent directly from the system, but it is recommended to also have access to the files in the directories on the server. An important place in addition to the admin panel is a panel on the server and access to ftp. New updates and documents can be sent directly from the system, but it is recommended to also have access to the files in the directories on the server.

The most popular CMSes like WordPress, Joomla or Drupal are based on GNU GPL license. It means [11]:

1. The freedom to run any program under this license, regardless of the purpose.
2. Freedom to analyze and modify the program for improvement.
3. Freedom to provide a program to help other users.
4. The freedom to improve the program and distribute their own modifications. It refers to the entire community.

In most cases, open source CMS means better quality [12]. The code is created by thousands of people [13].

The advantages of open source code pages:

1. Users do not pay for the system and add-ons available on the Internet. There is a large number of public and free graphic templates.
2. The code is updated regularly for the most popular systems like WordPress or Joomla. Thanks to the huge community of protection against attacks are high.
3. Easy migration between servers. The current standards of free CMSs allow simple data export and import if user move the site to another server on the network. It is just necessary to install the plugin and copy the generated files.
4. Intuitive interface developed over several years with thousands of users. Both administrative panels like front-end sites have a simple scheme that fits the standard user. Thanks to the appropriate layout, using the site does not require advanced programming knowledge.

The disadvantages of open source code pages:

1. Increased amount of spam, such as updates, occasional ads, system notifications from creators.
2. No custom applications. For individual processes and mechanisms, open-source systems may not include such functionality.
3. Not all platforms are up-to-date and supported by developers and communities. Only the most popular types of WordPress, Joomla, Drupal have full support and 24/7 support for users and administrators.
4. Users do not have manufacturer's warranty in case of hacking, other undesirable actions. In this case, administrator risks installing on his own responsibility and in the event of problems he need to contact the private companies supporting the systems or trust the community of the system.

The advantages of own content management platform:

1. Paid CMSs are more applicable to niche projects where standard solutions are not enough. For a reasonable fee, user will receive every product he desire.
2. High consistency of all modules. If the system is created by one programmer or one team, all elements are more consistent and fit together.
3. The creators of commercial solutions have an individual approach to the client. For larger companies, users get a product warranty and, in case of failure, immediate technical support and possible compensation. Users are confident that the developer will check the problem and help.

4. When buying an individual solution, users can ensure their uniqueness. By paying for an exclusive license, they are assured that the graphical interface or mechanism of action will not be used on any other website on the Internet.

The disadvantages of own content management platform:

1. The system user must pay for his purchase, often also for technical support and operation. Manufacturers also demand fees for additional extensions and plugins, as well as new graphical interfaces.
2. Managing such a system is often more difficult than free solutions, because it is based on copyrighted ideas and assumptions, is not up to standards.
3. Authorized systems are heavy in personalization. For CMS with open code, user can view available extensions and build a site by adding existing modules. In case of commercial applications, usually tools are written from scratch and the final effect can only be seen after the order has been fulfilled.
4. For small businesses, systems may not be fully secure, because of the many attack possibilities and a small number of IT supporters. It is necessary to make frequent copies and in the event of an attack, to secure the holes (learning errors more, compared to open source and thousands of programmers supporting the project).

5   The Usage of CMS as a Platform for City Dwellers

Content management systems such as WordPress, Joomla or Drupal, based on the GNU GPL are the ideal solutions for implementation of platform for city dwellers. Their flexibility and great functionality may be the root for the further development of the website, which will not necessarily cost a fortune in setting up and subsequent maintenance. Thousands of users around the world verify the safety and mistakes every time.

The author decided to divide the functionality of the portal on 10 spheres, which are integral with each other, but are responsible for other areas of everyday citizen. They are: smart business, smart human capital, smart management, smart environment, smart transportation, smart IT, smart everyday life & communication, smart care and smart future.

Each of them can be implemented in the portal so that the resident had a real impact on the operation of the city. The following is a discussion of each of them and the general scheme of the site. Every professional website must have the appropriate interface to their functionality and target the visitor. Most portals for content management like WordPress, Joomla or Drupal have a large possibility to modify the creation and appearance, and their functionality is virtually unlimited. It's important to customize every website for all users and target. The site should also be adapted for blind people and deaf. The appearance must be clear and simple, but at the same time modern. Important information concerning the accident or alert on important matters should be available in a place that is qualified to "method of 5 seconds" [14]. The following sections should be submitted to the above sections in the field of smart city, each in a different color (Fig. 5).

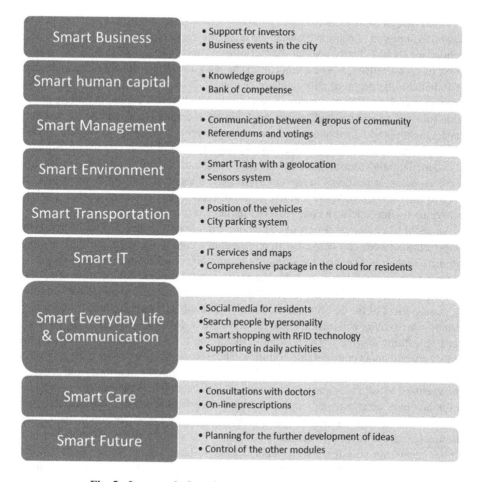

Fig. 5. Internet platform for city dwellers. Suggested modules.

Smart business - this is an area for residents engaged in business in the city, those who would like to review the offer opportunities to invest in the city, made cooperation with other cities and stakeholders. The interactive map should easily show locations of businesses to buy and invest. In addition to issues relating directly to the agglomeration should appear more universal elements for investors and residents themselves, such as trading companies and municipal rates. Users should be able to comment information from the business world. Another map should contain full information about the business events areas of the city. They can be added by the users and moderated by the staff offices. An important element that should be in the fold for entrepreneurs is counter debts. It is not designed for the same business, but also for the citizens. On the one hand, each user can monitor municipal finances, but on the other hand entities that want to invest can confirm the stability of the monetary authorities of management of agglomeration. There should be a special area for start-ups operating in the city.

Smart human capital has the task of improving the skills of citizens in all aspects of life and their proper allocation between the residents, universities and businesses. In this section, we should clearly present the full offer of universities and as well as create a "bank of competence", where users can exchange knowledge, information, aggregate inhabitants in the group of data fields in easy way. It is also valuable information for companies creating jobs in the city. Keep in mind that human capital is infinite added value for the city and the budget for development and increase should be large. Panel management should facilitate communication between citizens, civil servants, businesses and universities. Every citizen should have a direct influence on the decisions of city council through referendums and citizens' budgets. This can be easily solved thanks to the accounts of residents in the voting module integrated with the entire portal. In order to fully authenticate the identity of the residents, they can confirm their choices and voting by introducing a code received in the text message.

Smart Management is a part of the whole scheme that is responsible for the fully cooperation between city council with president of a city, universities, business entities and the social side. The tab should liquefy communication between these groups.

Smart Environment is a module that easily should provide information about the city. The author proposes here several innovative solutions that significantly simplify the management of the environment in the city. The tab will be considering smart trash system, which easily will inform about the pollution in the street. Thanks to geolocation, resident walking down the street can push the button on his mobile device, which tells the city council about considerable pollution. There is also taken a picture during sending the information. Officials using of these data estimate the amount of priority and generate a new request to the municipal services with a call to intervene. With time due to changes in the law may be chosen a company, which is responsible for the full process, from infrastructure through the information center at the city office, ending on managing the machines and employees of cleaning companies. A slightly different offer, but in a similar range can be offered to residents of single-family property, and government housing. By installing special sensors in the storage areas of waste can easily inform residents about the approaching overflow buckets and automatic call maintenance services. Residents with applications of Business Intelligence will be able to verify reports of fees and the number of call services.

Smart Transport is both public transport and private cars. Municipal vehicles should be promoted, because by reducing the amount of private cars decrease traffic jams and exhaust fumes. Solutions that can be delivered in the city dwellers platform in the most will be available in the mobile version - primarily geolocation module buses and trams. When you turn connection to the satellites you will be able to determine its current position and find the vehicles in real time on the selected filters which tram is the nearest looking at the current traffic, or whether it will be possible to change between the delayed bus and tram arriving as planned. Virtualization is very popular nowadays and can promote the popularization of urban transport. What more private drivers could use the portal page by synchronizing with sensors placed at parking in the whole city. Proper algorithm could determine the free parking spaces with an indication of the closest located from your current location.

The IT area allows residents access to the latest technologies. The portal could have a map of access points to the network, a list of the buses where travelers have possibilities to charge their electronics devices on the board, infographics that define the speed of the Internet in different areas in the city, elements informing about the technology provided by mobile networks and digital signal. As part of the account, each user could ultimately receive an e-mail account from the domain of the city, cloud storage for files and ftp account with databases to create a business card in the Internet. Smart IT should support both ordinary citizens in their daily duties and activities as well as entrepreneurs focus their network and promote outside the agglomeration.

Smart everyday life & communication is combined into one tab because interpersonal communication is an integral part of everyday life and it is essential that all these processes were performed. The website would also comply with a social function, so people in the area could get to know each other, organize joint events, assist in discussion groups and forums and develop a community network, which will be the nucleus of the real society. Portal could segregate people by profession, age, common interests. It will be support to finding the staff to companies, friends to the cinema or to organize a thematic meeting. Smart shopping module could help shopping in the store. In principle, residents would get a magnetic card, which can be an integrated part of their wallet. All the goods in the store would be equipped with a RFID card, which after crossing the gate at the exit of the store to read and lent to an electronic card, which is directly linked to a bank account. Notification of mobile application will be sending a request whether the resident made shopping for that amount and will be required to identify him via a password. In the case of successful verification and collection of money from the account. Next, the exit door is automatically open and the resident may leave the store. The aim of these measures is to reduce the number of "smart" solutions in the one place and create an universal interface to perform many tasks.

Smart care is a module responsible for the health of residents. Each patient through the portal will be able to book a visit with a doctor to verify results of test, check the status of prescription and order pharmaceuticals online. By installing special medical equipment portal will be able to send reports on the state of health of the doctors.

Smart Future is the last area responsible for the development of the city and the idea of a smart city. It will monitor the remaining 9 modules and show the further development directions. Residents will be able to conduct their own studies and projects within the smart city, and the tab will promote and popularize the most interesting ones.

6   Conclusion

The creation of the Internet Platform for city dwellers, which includes 10 of mentioned above areas based on content management systems under the GNU GPL open a new path in the development of smart city. The combination of all these aspects in one place will increase the efficiency of internal processes and improve the exchange of information. It is also important to standardize all schemes, systems and applications, which work in the field of the platform. Each city is guided by a different specificity and has other conditions, however part of the functionality is shared by all. It is worth considering creating one central core of the system for several cities in a nearby location and designing dedicated solutions for each locality. Content management systems based on open source have the ability to maintain the platform for dwellers. Unfortunately, currently, despite the many thousands of add-ons and extensions there is no clearly defined path of development systems in terms of smart city. This also applies to the business path that can take place when developing urban solutions using open source as well as scientific. There is no literature in the area that specifies the development of content management systems in relation to smart city.

The article was intended to familiarize the issues related to the use of systems with virtualization implementations of the city, as well as suggest the examples of innovative solutions which will enhance the safety and quality of life of every citizen. The author believes that the popularization of content management systems based on open source code in relation to smart city issues may in the future create solutions to help human development. Year after year, we are witnessing a growing computerization of society and the problem of humanity moving from rural to urban areas. In many cases, it may be the implementation of information systems that will monitor many factors in the city, reduce traffic jams, facilitate access to medical care for residents, and help report incidents.

References

1. Komninos, N.: Intelligent Cities and Globalization of Innovation Networks, p. 17 (2008)
2. Chourabi, H., Nam, T., Walker, S., Gil-Garcia, J.R., Mellouli, S., Nahon, K., Pardo, T.A., Scholl, H.J.: Understanding smart cities: an integrative framework (2012)
3. Townsend, A.M.: Smart Cities: Big Data, Civic Hackers, and the Quest for a New Utopia, p. 93 (2013)
4. Dobbs, R., et al.: Urban world: cities and the rise of the consuming class, McKinsey Global Institute (2012)
5. Mercier-Laurent, E.: What technology for efficient support of sustainable development (2015)
6. Nam, T., Pardo, T.A.: Conceptualizing smart city with dimensions of technology, people, and institutions (2011)
7. http://www.edinteractive.co.uk/article/?id=4
8. http://www.seguetech.com/blog/2013/05/01/client-side-server-side-code-difference
9. http://www.techterms.com/definition/cms
10. Mehta, N.: Choosing an Open Source CMS. Beginner's Guide, p. 19. Packt Publishing, Birmingham (2009)
11. Lindberg, V.: Intellectual Property and Open Source, pp. 341–354 (2008)
12. Weber, S.: The Success of Open Source, p. 54 (2004)
13. Connor, C.: How to Create a Website for Business or Personal Use, p. 17 (2013)
14. Johnsen, M.: Multilingual Digital Marketing: Become the Market Leader, p. 26 (2016)

Segmentation of Social Network Users in Turkey

Mahmut Ali Özkuran[1]([⊠]) and Gülgün Kayakutlu[2]

[1] Istanbul Technical University, Istanbul, Turkey
ozkuran@gmail.com
[2] Istanbul Technical University, Industrial Engineering Department, Istanbul, Turkey

Abstract. In the digital world that we are living on companies needs as much as knowledge as they can collect to survive. Knowledge creation process requires as much as data to create more dimension on the knowledge. As the knowledge created with data in different dimensions it become more useful for the company. At this point, Knowledge Management helps creation of additional information sources to be used for knowledge accumulation.

In today's digital world social networks are fruitful data sources for nearly all of industries. Companies that utilizes this data source easily add another dimension to their information base and creates more revealing knowledge to their Knowledge Base. With automated processes, up-to-date information can be added to the Knowledge Base of the companies with well-known, easy to apply methods with little effort.

In this paper, we have conducted a research to reveal segments of the Turkish Twitter users using Self Organizing Maps method. Results show that, using segmentation we can create an important knowledge source about focus of interest, which could be used as a tool for analyzing the market penetration of the advertisements.

Keywords: Segmentation · Social networks · Knowledge management
Self-organizing maps

1 Introduction

We are living in the knowledge era. Everything around us from smartphones to "Internet of Things" devices creates uncountable amount of data. According to IDC, size of the digital universe is doubling every two-year [1]. Without organization of that much data, it is not possible to create information and knowledge that will help to create better processes for the companies. Knowledge Management offers different tools to convert this data for better utilization for the company. Those knowledge creation tools make Knowledge Management as one of the most important processes for the modern businesses.

Smart companies convert every step of their processes to the data, and using this data they create information and knowledge about those processes. Created knowledge and information helps with the optimization of every step of the company's commercial

© IFIP International Federation for Information Processing 2018
Published by Springer International Publishing AG 2018. All Rights Reserved
E. Mercier-Laurent and D. Boulanger (Eds.): AI4KM 2016, IFIP AICT 518, pp. 119–131, 2018.
https://doi.org/10.1007/978-3-319-92928-6_8

effort from production to marketing. Especially optimization of the marketing opportunities is limitless if data of the customers can be converted into knowledge.

Usage of direct customer information collecting methods like Marketing Research (Qualitative or Quantitative) and Customer Relation Management have presents different dimensions of the customer knowledge. Adding new dimensions to this knowledge base would be help with the better understanding of the customers' needs and requirements.

As we are under the rain of data today, we have many different possibilities for adding this new dimension. As it offers relatively cheap data that varies over all segments of the society, social networks are very suitable tool for adding this new dimension.

From advertising to customer relation management, social networks are offering direct communication chance between customers and companies. Smart companies are utilizing this direct communication chance with their customers and make their products and processes better using those feedbacks. These optimizations made in the process and products also helps with the increases of their market share. In addition to the direct communication opportunity as an ocean of information, social networks give hints about many different properties of their users. Using these properties, also helps with selecting right audience for the targeted advertisement and opportunity of more concentrated connection with potential or existing customers of the products.

As we mentioned above social networks are widely used by any type of consumer, independent of revenue or belief. This variety in users makes social networks one of the important sources to discover potential markets. As there are many to learn from the social networks we should start with classification of the users of those networks and evaluate the potential marketing opportunities.

Some example research already made on social networks as an example of their productivity on creation of information. A sample work done by García-Palomares et al. shows some possible knowledge that can be learned from the social networks [2]. Researchers are classified cities using the GIS tags of photos shared on the social networks. According to the research every European metropolis have different tourist attraction. As some of the cities like Paris and London have dispersion in their touristic photo hotspots, other cities like Athens and Rotterdam has concentration of photos on some spots.

Today there are thousands of different sized social networks that servicing internet users. Twitter as one of the biggest social networks, (320 Million active users [3]) its data creates many possible areas for researchers to work on. From usage density of the certain words or tags to tweet count by time on any special event, their data is available to researchers. We have used this data source to find followers of the Turkish newspapers and other well-known Twitter accounts to classify Turkish twitter users by using segmentation.

The article organized that the next section is reserved for the literature review and the third section for the method. The application and results will be analyzed in the fourth section and conclusion will be given in the last section.

2   Literature Review

Many aspects of our lives have been digitized in the information era. This digitized data opened many avenues for scientific research. Especially public domain data that created by service users let researchers to created many new methods to work with.

2.1   Social Networks in Research

The Social network data presents limitless possibilities for the scientists to conduct a research on different fields from sociology to market research. For example, social networks are well known for their information exchange speed. This speed in information propagation inspires companies to create social networking tools that let their employees to share their knowledge. Behringer and Sassenberg made a cross-sectional study on the effects of different features on using social media as knowledge exchange [4]. But resistance against those new tools among the employees also affected the social media knowledge sharing applications.

Social networks are not only good for knowledge sharing but creating the knowledge. Companies are using social networks as data source for their strategic plans. Nguyen et al. conducted a research uses Chinese online sector data to examine the relationships between the effect of knowledge acquisition from social media, proactive and reactive market orientation, strategic capability over social media and brand innovation strategy [5].

Usage of social networks are not only important for big companies but also important for small and middle sized enterprises [6] and individuals. Social networks are not only diffuse information between the users but in their interests too. A research made by D'Agostino et al. showed that parts that constitutes a social network is effected by its neighbors and trends [7].

Reciprocity is an extension of human behavior into the social networks. According to social exchange theory social behavior of a person is the result of an exchange process. Maximizing the benefit with minimized cost is the purpose of the social exchange. At this point we can say that online social networks are perfect for maximizing the profit with less effort. Reciprocity has important part of social exchanges as humans are tends to keep score, increase in the number of the reciprocity messages from the user, increases the number of reciprocity messages from his/her audience [8]. Usage of social networks is also related to other aspects human behavior like extroversion, conscientiousness, agreeableness, openness and emotional stability [9–11].

As behavior of individual social network users important to discover the dynamics of social networks they are still made with small samples compared to the times of Big Data. Today data of huge networks are used to find out the habits of users [12]. Not only the habits of users, also relations between users are important source of information. People uses micro blogging sites for following persons with similar interests or sources of information that interested in. This relation data between the users can be used for knowledge creation [13].

3   Methodology

In the fast times that we are living on, every information that created has an expiration date. To overcome information has propagation speed that never heard of before, we need to get use of information as fast possible to create difference from our competitors. With this background given we can say that it is important to create information from real time data with well-known data processing methodologies and automated processes.

3.1   Segmentation

Increase of the digital technologies, number of channels between customers and companies increased unprecedentedly. This increase in number of channels enforces companies to produce different marketing campaigns and products to reach out that many different markets. At this point segmentation of the customers (or even potential customers) become important for today's companies. Segmenting markets gives companies opportunity to create different marketing campaigns and products for different segments of the market.

A research made by Hamka et al. in 2014 used segmentation to differentiate mobile services users [14]. Classical segmentation of mobile device users is generally based on location of the calls or the length of the calls, but in this research, scientists gone one step further and measured the behaviors of users using a software. Researchers found that mobile service users can be segmented by usage of the network and the usage of the content services.

3.2   Self-organizing Maps

Applications of social networking especially in the problems of coordination and cooperation make social network analysis embedded into many research fields. In any analysis, easiest representation of social networks is representing them as vertices (elements) and edges (relations). Prior to 1970's presentation of social network are made by sociograms [15] and modern graph theory [16]. From the 1970's, as data and processing power of the computers increased analysis and research of social networks has become hot topic.

Among all those advances in social networks research, Finnish professor Teuvo Kohonen introduced the Self-organizing maps in 1980s [17]. Self-organizing map is an artificial neural network that converts some input into a map (usually two dimensional) using unsupervised learning. Those maps are consisting of points called neurons. As name implies working model of neurons is inspired by the human brain data handling method.

At start weights neurons are initialized with random values. Following the initialization, competitive learning starts with supplying of test data consists of vectors to the network. With each new training vector supplied to the network, distance to all weight vectors is calculated and the most similar neuron to this input is called best matching unit. The best matching unit and neighbors of this neuron come closer to the given vector.

As supplied training vector count increases this adjustment amount toward given input decreases. Function that updating the neurons is given at formula (1).

$$W_v(s + 1) = W_v(s) + \alpha(s)(\theta(u, v, s)(T - W_v(s)) \tag{1}$$

Where **s** iteration number, **T** training vector, **W** nodes weight vector, **u** index of best matching neuron, **v** index of the neuron to be calculated, **θ(u,v,s)** distance function from best matching neuron to neuron **v** and **α(s)** as training function makes decreases the effect of training vector as supplied data count increases.

As it is a powerful tool for converting multidimensional data in to two dimensional maps self-organizing maps have wide usage area in different topics. Kohonen et al. mentioned about those different usage areas in their paper dated back to 1996. According to this paper, self-organizing maps can be used for feature detection in signal processing, fault detection in process and system analysis, in pattern recognition or even in robotics for navigation [18].

An example research by Bhandarkar et al. gives shows how to use self-organizing maps on image segmentation [19]. Researchers are used an improved version of the original self-organizing map, the hierarchical self-organizing map. This hierarchical self-organizing map is makes application of vector quantization on images possible, which leads to segmentation of the images.

Self-organizing map is also used in many researches related to market segmentation. A research by Bloom dated back to 2004 made market segmentation for the tourists using self-organizing map and backpropagation neural networks [20]. Bloom visualized the segmentation of the tourists visiting South Africa.

A recent research prepared from Iranian ADSL subscriber data also shows implementation of self-organizing map onto market segmentation [21]. The researchers are integrated Fuzzy Delphi method and self-organizing maps to visualize segments of the market.

4 Application and Results

We have made our research on Twitter data. Using Twitter API, we have downloaded IDs the followers of 74 selected Twitter accounts. Twitter accounts of selected Turkish Newspapers, Universities, Art Museums, Online Learning Platforms and Actors as initial research data. We have downloaded those IDs using C# and processed this data using Python and applied self-organizing map algorithms using R. SOM function from Kohonen package of R used for creating self-organizing maps.

Initially we have created main data consists of followers of 22 twitter account that belongs to 18 newspapers, 2 online newspapers and 2 television channels. Furthermore, follower data of those newspaper accounts mixed up with selected accounts related to Education, Art and Cinema. Combined data used for creating different Self-Organizing Maps.

To simulate creation of different datasets for different information creation process we have combined different Twitter accounts to create 5 different combinations of data.

- Dataset 1 consists of the followers of top 5 bestselling Turkish newspapers Twitter accounts.
- Dataset 2 consists of the followers of the all Turkish newspapers Twitter accounts.
- Dataset 3 consists of the followers of the all Turkish newspapers and art related Twitter accounts
- Dataset 4 consists of the followers of the all Turkish newspapers and Cinema related Twitter accounts
- Dataset 5 consists of the followers of Cumhuriyet, Hurriyet, Radikal newspapers and well known universities Twitter accounts.

Application of self-organizing maps on to data has created datasets in size of 30 by 30. For each dataset, Self-Organizing Map algorithm is fits after 60 iterations as seen in Fig. 1.

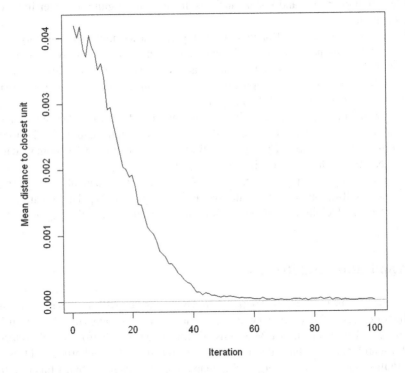

Fig. 1. Training progress

For visualizing results of the self-organizing map method data, we have created heat maps, neighboring distance maps, counts plot and cluster maps using those datasets. General overview of visual results of research on Dataset 1 is shown in Fig. 2.

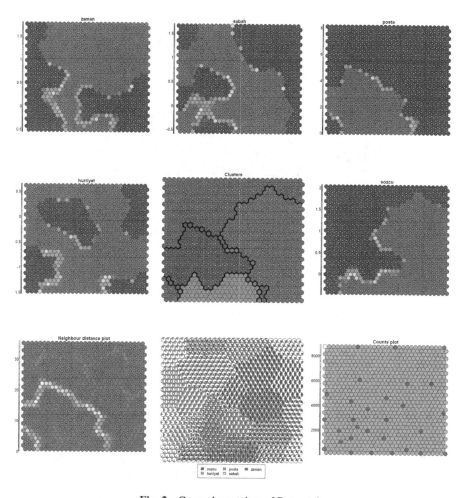

Fig. 2. General overview of Dataset 1

Application of self-organizing map algorithm on to Dataset 1 undercover 4 big segments which are visualized in Fig. 3.

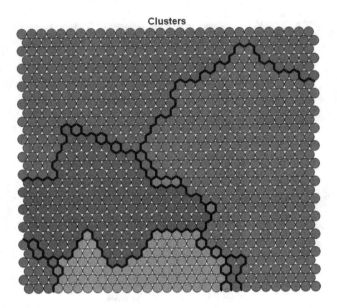

Fig. 3. Clusters created by application of SOM on Dataset 1

Followers of twitter accounts of mainstream newspapers like Hürriyet, Milliyet and Habertürk is scattered all around the map without creating big chunks of segments (Fig. 4).

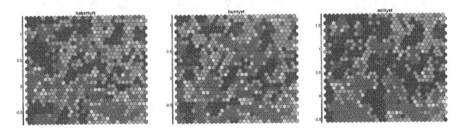

Fig. 4. Dispersion of mainstream paper followers

Followers of twitter accounts of newspapers Akşam, Sabah, Vatan, Yeni Şafak, Star, Zaman and Habertürk are overlapping while Sabah and Zaman have broader follower base (Fig. 5).

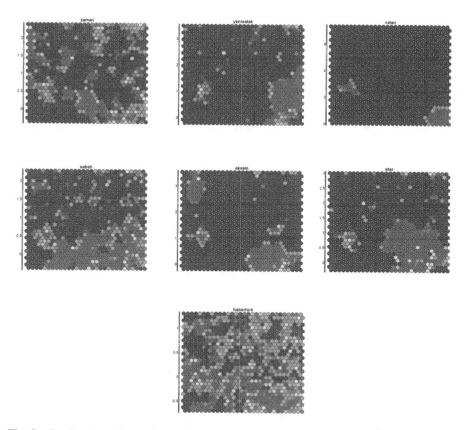

Fig. 5. Overlapping follower bases of similar media companies - Akşam, Sabah, Vatan, Yeni Şafak, Star, Zaman and Habertürk.

Followers of twitter accounts of newspapers Aydınlık, Cumhuriyet, Diken, Sözcü, Halk TV, Sol and Ulusal Kanal are overlapping while Cumhuriyet and Sözcü have broader follower base (Fig. 6).

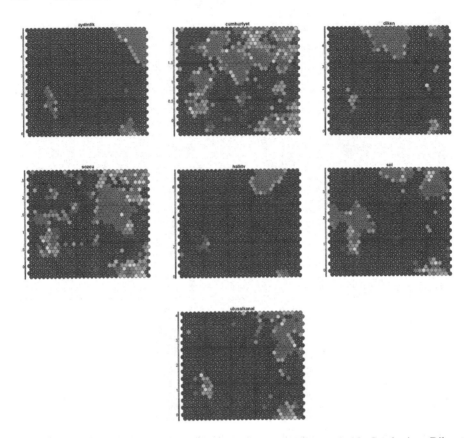

Fig. 6. Overlapping follower bases of similar media companies - Aydınlık, Cumhuriyet, Diken, Sözcü, Halk TV, Sol and Ulusal Kanal.

Radikal and Cumhuriyet twitter accounts have similar follower base while Radikal follower more packed up (Fig. 7).

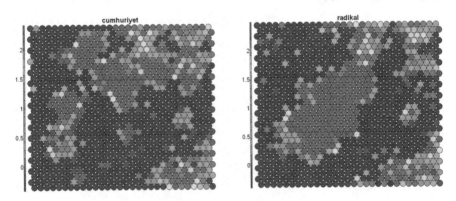

Fig. 7. Followers of twitter accounts of Cumhuriyet and Radikal

Followers of twitter accounts of Universities are overlapping the followers of Radikal (Fig. 8).

Fig. 8. Followers of Radikal, Hurriyet and University twitter accounts

Followers of twitter accounts of Cinema related accounts are overlapping the followers of Radikal.

Results shows followers of the twitter account belong to Radikal newspaper are have a remarkable interest in universities and cinema. Consequently, Radikal newspaper seems have a good follower base for adverts of educational institutions and films.

For more solid forecasting for successful advertisement campaign we may use the numerical results of the self-organizing map method.

5   Conclusion

Social Networks are abundant data sources for researchers. Especially in the market research domain conducting researches using this data source can go beyond the classical scope of the market research. Researchers can identify the requirements of the certain segments of market using the knowledge created using information that distilled from social network data.

There are many algorithms and research methods are available to find the segments of any data. In this research, we have selected the self-organizing map method to find the segments of the social network users and visualize them. Application of self-organizing maps on social network data is productive in the means of knowledge creation. Using this data, we have created knowledge about general habits of followers of different Twitter accounts. For example, Turkish followers of the Twitter accounts of the universities are overlapping with followers of the Radikal Newspaper.

This research method also allows automated data creation, which may lead to real-time information retrieval. As we analyze the process of this research, it is possible to get real-time follower data from Twitter API to our data warehouse. Using data in this data warehouse we may run of segmentation scripts continuously and we may get up-to-date information about the segmentation of Twitter users with given inputs.

Several methods can be compared for virtual marketing; or analysis can be enriched by industry based specific subjects. Pointing the target audience is very important for both getting the right result and more effective marketing campaign. Using ids of the users already present in this data we may create tailor made offers for each and every customer.

For future research, we may improve those results with different research techniques like text mining and surveying for creating even more segments of Twitter users. Creation of time series that shows differences in segmentation could be also a good research opportunity. Also, our research resulted interesting group of individuals around the maps, researching those individuals with surveys may reveal focus segments.

References

1. Turner, V., Gantz, J.F., Reinsel, D., Minton, S.: The Digital Universe of Opportunities: Rich Data and Increasing Value of the Internet of Things. IDC White Paper (2014). https://www.emc.com/leadership/digital-universe/2014iview/index.htm?cmp=micro
2. García-Palomares, J.C., Gutiérrez, J., Mínguez, C.: Identification of tourist hot spots based on social networks: a comparative analysis of European metropolises using photo-sharing services and GIS. Appl. Geogr. **63**, 408–417 (2015)
3. Twitter: Twitter | About (2016). https://about.twitter.com/company. Accessed 03 Mar 2016
4. Behringer, N., Sassenberg, K.: Introducing social media for knowledge management: determinants of employees' intentions to adopt new tools. Comput. Hum. Behav. **48**, 290–296 (2015)
5. Nguyen, B., Yu, X., Melewar, T.C., Chen, J.: Brand innovation and social media: knowledge acquisition from social media, market orientation, and the moderating role of social media strategic capability. Ind. Mark. Manag. **51**, 11–25 (2015)
6. Fernandes, S., Belo, A., Castela, G.: Social network enterprise behaviors and patterns in SMEs: lessons from a Portuguese local community centered around the tourism industry. Technol. Soc. **44**, 15–22 (2016)
7. D'Agostino, G., D'Antonio, F., De Nicola, A., Tucci, S.: Interests diffusion in social networks. Phys. A Stat. Mech. Appl. **436**, 443–461 (2015)
8. Surma, J.: Social exchange in online social networks. the reciprocity phenomenon on Facebook. Comput. Commun. **73**, 342–346 (2016)
9. Ross, C., Orr, E.S., Sisic, M., Arseneault, J.M., Simmering, M.G., Orr, R.R.: Personality and motivations associated with Facebook use. Comput. Hum. Behav. **25**(2), 578–586 (2009)
10. Amichai-Hamburger, Y., Vinitzky, G.: Social network use and personality. Comput. Hum. Behav. **26**(6), 1289–1295 (2010)
11. Moore, K., McElroy, J.C.: The influence of personality on Facebook usage, wall postings, and regret. Comput. Hum. Behav. **28**(1), 267–274 (2012)
12. Wang, Z., Tu, L., Guo, Z., Yang, L.T., Huang, B.: Analysis of user behaviors by mining large network data sets. Future Gener. Comput. Syst. **37**, 429–437 (2014)
13. Liang, B., Liu, Y., Zhang, M., Ma, S., Ru, L., Zhang, K.: Searching for people to follow in social networks. Expert Syst. Appl. **41**(16), 7455–7465 (2014)
14. Hamka, F., Bouwman, H., De Reuver, M., Kroesen, M.: Mobile customer segmentation based on smartphone measurement. Telemat. Inform. **31**(2), 220–227 (2014)
15. Moreno, J.L.: Who Shall Survive? A New Approach to the Problem of Human Interrelations, vol. 58 (1934)
16. Harary, F., Norman, R.Z., Cartwright, D.: Structural Models: An Introduction to the Theory of Directed Graphs. Wiley, New York (1965)
17. Kohonen, T.: Self-organized formation of topologically correct feature maps. Biol. Cybern. **43**(1), 59–69 (1982)

18. Kohonen, T., Oja, E., Simula, O., Visa, A., Kangas, J.: Engineering applications of the self-organizing map. Proc. IEEE **84**(10), 1358–1383 (1996)
19. Bhandarkar, S.M., Koh, J., Suk, M.: Multiscale image segmentation using a hierarchical self-organizing map. Neurocomputing **14**(3), 241–272 (1997)
20. Bloom, J.Z.: Tourist market segmentation with linear and non-linear techniques. Tour. Manag. **25**(6), 723–733 (2004)
21. Hanafizadeh, P., Mirzazadeh, M.: Visualizing market segmentation using self-organizing maps and Fuzzy Delphi method - ADSL market of a telecommunication company. Expert Syst. Appl. **38**(1), 198–205 (2011)

Towards Semantic Reasoning
in Knowledge Management Systems

Gulnar Mehdi[1,2]([✉]), Sebastian Brandt[2], Mikhail Roshchin[2],
and Thomas Runkler[1,2]

[1] Siemens Corporate Technology, Munich, Germany
{gulnar.mehdi,thomas.runkler}@siemens.com
[2] Technical University of Munich, Munich, Germany

Abstract. Modern applications of AI systems rely on their ability to
acquire, represent and process expert knowledge for problem-solving
and reasoning. Consequently, there has been significant interest in both
industry and academia to establish advanced knowledge management
(KM) systems, promoting the effective use of knowledge. In this paper,
we examine the requirements and limitations of current commercial KM
systems and propose a new approach to semantic reasoning supporting
Big Data access, analytics, reporting and automation related tasks. We
also provide comparative analysis of how state-of-the-art KM systems
can benefit from semantics by illustrating examples from the life-sciences
and industry. Lastly, we present results of our semantic-based analytics
workflow implemented for Siemens power generation plants.

Keywords: Knowledge management · Semantic technology
Data-access · Analytics · Automation

1 Introduction

It is well established today that knowledge is the core element of any AI based
system, be it small robots like Roomba [1] or large-scale applications such as IBM
Watson [2]. Consequently, the transition of the global economy towards knowledge economy is an evident and prominent process in our information society.
Even small-scale industries today value knowledge resources and use them for
gaining a competitive edge. From a research and technology view point, much
progress has been made in enabling information systems to leverage knowledge
for decision-making and analysis. The scope of these systems is to construct,
manage, share and process the applicable knowledge for their respective tasks.
For example, KM systems are build to not only manage large repositories of
biomedical data coming in from lab reports, patient records, research papers,

This research is supported by the Optique project with the grant agreement FP7-
318338.

E. Mercier-Laurent and D. Boulanger (Eds.): AI4KM 2016, IFIP AICT 518, pp. 132–146, 2018.
https://doi.org/10.1007/978-3-319-92928-6_9

and medical imaging, but also to store and analyze useful patterns and knowledge from them [3]. Nevertheless, implementing such a KM system is complex and incorporates multi-faceted concepts from various disciplines and business practices. For instance, information scientists consider taxonomies, subject headings, and classification schemes to represent knowledge, whereas consulting firms actively promote practices and methodologies to capture corporate knowledge assets and organizational memory. In the biomedical industry, knowledge management practices often need to leverage existing clinical decision support, information retrieval, and digital library techniques to capture and deliver tacit and explicit biomedical knowledge [3]. Engineers, on the other hand rely on knowledge and data-driven strategies for design, manufacturing and maintenance of their artifacts. According to KPMG and the Conference Board [4], 80% of the world's biggest companies have knowledge management efforts under way, especially in the medical domain. Nevertheless, the full potential of KM systems is yet to be unlocked. Specifically, the challenges with respect to knowledge representation, search, integration, data-access and reasoning etc. are well understood in the research community but existing solutions have rarely resulted in widely adopted practical implementation. In this paper, we discuss requirements and limitations of the existing semantic and non-semantic solutions for KM and propose a new approach to semantic-based knowledge management that does not only enhance the feature set and usability but also supports analytics, reporting workflows, automation and big data infrastructures. In Sect. 2, we present the related approaches and further discuss ontology-based KM systems in Sect. 3. Sections 4 and 5 presents success stories from the healthcare and engineering domain along with their extended requirement set. Section 6 describes the current challenges for semantic approaches and in Sect. 7 we propose our solution along with the results from Siemens Turbo-machinery use-case.

2   Related Approaches

Building and using KM system involves many tasks, see Fig. 1. First and foremost is knowledge acquisition and representation, into which the scientific community has invested much time and effort. Knowledge engineering [5] methodologies for building expert systems have applied knowledge acquisition techniques (e.g. interviewing, protocol analysis, simulation, personal construct theory, card sorting, etc.) for eliciting the tacit knowledge from domain experts. Knowledge acquisition techniques are applied in order to develop knowledge repositories in knowledge management systems for formally documenting knowledge in a machine-processable way. To represent knowledge, a knowledge taxonomy and knowledge mapping are typically constructed for serving as a framework for building knowledge repositories [5]. Ontologies and ways for representing acquired knowledge (rules, cases, scripts, frames/objects, semantic networks, etc.) are typically created in the AI field for building expert and other intelligent systems [6]. Natural language and speech understanding front-ends as interfaces to knowledge management systems are important additions to enhance search over and dissemination of knowledge. Data mining and knowledge discovery techniques are

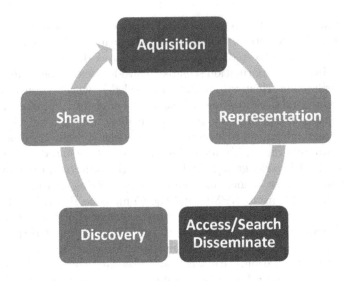

Fig. 1. Different facets of knowledge management systems

employed to inductively look for trends, relationships, clusters and possibly new insights and information from knowledge repositories [8]. Online communities with a common interest in knowledge management are ways of sharing and distributing knowledge. Intelligent agents on the other hand are also applied to analyze the knowledge, email, web pages, and the like and to disseminate appropriate summaries or individual pieces of information and knowledge to those who should best make use of it [7].

Limitations. Most previous works on KM systems has focused on its success factors [9–11]. There are a few studies on challenges or limitations to KM systems [12–14]. However, they often offer intrinsic business value but KM systems do not always improve organizational performance because there exist some discrepancies between innovation and performance. We classify the requirements and limitations into two types: technological factors and social/cultural factors involving people [15]. Following are the significant weaknesses of current KM systems:

(i) **Searching often ignores context** in current KM system. Applications today provide key-word based search functionalities that often retrieve irrelevant information when terms have different meaning in different context or fail to relate different pieces of information into a meaningful context.

(ii) **Inefficient access and integration of information** is a major challenge in current systems. Human browsing and reading is required to extract and integrate information from different sources. Existing KM systems rely on labor intensive extract-transform-load jobs because the automatic agents do not integrate and possess common sense knowledge required to extract information from heterogeneous sources.

(iii) **Maintaining knowledge** is the main pain points of the state-of-the-art. It becomes difficult and time-consuming activity when the knowledge repositories become large or reach the level of Big Data infrastructures, for instance, hadoop clusters, teradata warehouse etc. Existing solutions lack transparency and usability with domain oriented interfaces as well as find it difficult to keep the knowledge consistent, correct and up-to-date. If the so called 'grain size' of the knowledge representation is chosen properly (i.e. small enough to be comprehensible but large enough to be meaningful to the domain expert) then the KM system will allow great flexibility for adding, removing or changing as well as querying knowledge in the system.

(iv) **Generation of documents and reports** is too cumbersome and slow process to execute in any industrial setting. Whereas, automated and context-aware authoring can be of greater advantage in enabling content according to user profile or other aspects of relevance. The generation of such information presentation would require machine-accessible representations of the semantics of the information sources.

(v) **Lack of shared understanding**, is another aspect which involves how people, organization and KM systems communicate with one another. It is obvious that each stakeholder has different needs and background context.

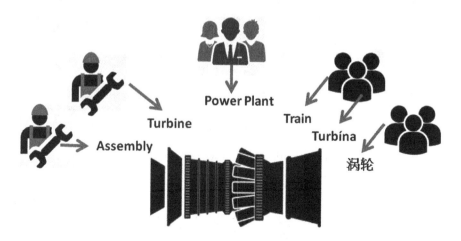

Fig. 2. Different naming but same semantics

Figure 2 represents a real-life example where a diverse set of users use different jargon in different languages to define the same concept or subject matter, that is Siemens Gas turbine in this case. This lack of shared understanding leads to disparate modelling methods, paradigm, languages and software tool which ultimately limits interoperability, reuse and sharing.

3   Semantic-Based Knowledge Management System

The vision of semantic technologies is to provide human-readable artifacts annotated with meta-information. This meta-information defines what the artifact is about in a machine-processable way. Ontologies are at the core of semantic technologies. Ontology is a formal explicit description of concepts, relations and properties of a domain. Knowledge in ontologies can be formalised using five kinds of components: classes, relations, functions, axioms and instances [17]. The meta-information together with domain ontologies provides an arena of knowledge-driven systems and automated services such as information access, reasoning services etc. It facilitates knowledge sharing and re-use and offers a wide feature set to support knowledge management capabilities. Figure 3 describes the unifying semantic architecture where ontologies are used throughout the knowledge management life cycle.

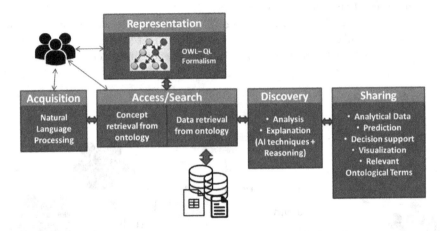

Fig. 3. Semantic solution to knowledge management systems

Ontologies are populated in an automated or semi-automated fashion from heterogeneous data sources. This serves the purpose of knowledge acquisition. This utility helps users to relate or map their domain concepts with the data model underneath and extract relevant information when required without performing cumbersome ETL jobs. As knowledge is acquired, it is then represented in an ontology language. The ontology language (for example: ontology web language - OWL, RDF etc.) is able to represent domain knowledge with a clear semantics as well as to provide redundancy or inconsistency checks. Knowledge maintenance is simplified by following standard design and modular approach to build and manage semantic models. Finally, the knowledge is made available to the end-user by means of semantic-based search, sharing, summarizing, visualization and organization.

3.1   Benefits of Using Ontologies

A key advantage of ontologies over many other knowledge representation formalisms is their formally well-defined semantics. They specifically support subsumption relations, multiple view points and hierarchies including partonomies, and inferred relationships. Ontology inference engines are used to derive implicit knowledge from explicit statements, detect redundancies and inconsistencies, and discover relationships that may not have been clear to the author of the ontology in the first place [17]. Currently, the most commonly used ontology formalism is OWL and its sub-languages. Some additional characteristics of ontology [22] addressing key challenges in the KM domain are:

- Ontologies clarifies the structure of knowledge and domain for an effective KM system.
- They separate factual knowledge about the domain from problem-solving knowledge.
- They facilitate sharing and re-using knowledge as well as interoperability of information resources between humans and software agents.
- They make searching, querying and browsing information more effective. For instance, a web site or a corporate intranet can organise its content according to some ontology which then can be utilised to improve the quality of searches. For instance, generalisation or specialisation of information can help in assisting users. It is a short distance from general search applications to knowledge management applications. One of the big challenges in knowledge management is to find knowledge and information that is relevant, and here ontologies have a lot of potential.
- They promote ease of maintenance of knowledge models and artefacts in KM system. The unified semantic solution can keep the information up-to-date with minimum extra effort, and the link between different types of knowledge can be examined by means of automated well-defined procedures.
- They provide a layer of abstraction over KM system services and are able to integrate heterogeneous knowledge resources. Various applications of ontology-based data access and integration are success stories in many industrial use-cases.

4   Healthcare Use-Case

Expert systems in the healthcare domain dates back to the early 70s, when the MYCIN program was developed to support consultation and decision-making. This expert system relied on expert knowledge in form of IF-THEN rules. Creating and encoding these rules was a time-consuming and labor-intensive process [3]. Later, medical terminologies were represented as ontologies along with many AI techniques such as data-mining, text mining, natural language processing etc. A prominent development is the construction of SNOMED CT and the gene ontology. SNOMED CT is a systematic collection of medical terms including codes, synonyms and definitions used in clinical documentation and reporting [18]. It also includes: clinical findings, symptoms, diagnoses, procedures, body structures, organisms and other ethologies, substances, pharmaceuticals, devices and specimens. Such a knowledge model facilitates sharing and aggregation of patient records and findings, the use of standards, access to heterogeneous data, ensures quality screening, facilitates treatment and electronic recording.

Figure 4 gives an example of lung disease and its causative agent in SNOMED CT. The example combines is-a and attribute relationships. Such a representation of terms in SNOMED ontology helps users in performing subsumption that is testing pairs of expressions to see whether one is a subtype of the other and vice versa as well as classification that is to structure a set of expressions according to their subsumption relationships.

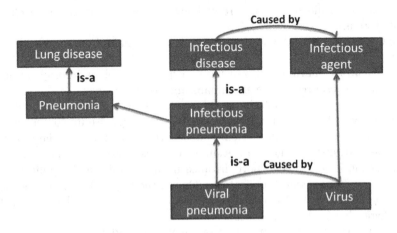

Fig. 4. SNOMED ontology example

Existing solutions in the healthcare focus on reference terminology applications where reasoning is often hidden behind the terminological efforts. Though systems without semantic models would find difficult to detect and repair inconsistencies in their knowledge repositories. Organizations such as IHTSDO, WHO etc. provide standard terminologies to be used by the industry and thus the healthcare domain is not much affected by knowledge authoring problems.

5   Engineering Use-Case

The engineering domain is well-defined and deals with known concepts and relations. Nevertheless, KM systems in the engineering domain often provide limited capabilities of search, data-access, integration and analytics [19–21]. The prime focus of many manufacturing and service industries today is to adopt data-driven strategies where they aim to move analytics and decision-support to the data itself. These strategies involve a variety of tasks from data-access, to integration, to storage, analytics, reporting and automation. Thus, a task such as data-access does not live in isolation anymore. Information systems are required to provide workflows with clear semantics to support these strategies and enable autonomous systems. For example, automatic shut-down of a power plant in case safety checks are violated.

It is important to realize the utility of semantic reasoning and the benefits it brings to the engineering applications. Current state-of-the-art applications of data integration, search and interoperability either use manual intervention of experts performing complex Extract-Transform-Load jobs or involve managing large set of configuration files. All these solutions require greater expertise and consume much time and effort. The existing implementations also lack automated reasoning capabilities because most of them are not based on logic-based formalisms. Semantic reasoning can provide better knowledge management services. It promotes reuse of models including wide range of ontologies such as standard sensor network (SSN) ontology[1] to address the domain requirements.

Figure 5 shows a snapshot of ontology developed for industrial gas and steam turbines. It captures concepts related to compositional structure of the plant, its processes and configurations. This semantic model helps engineers to infer relationships about plant configurations, processes and supports multiple hierarchies to represent part-of and is-a relationships of the related physical entities.

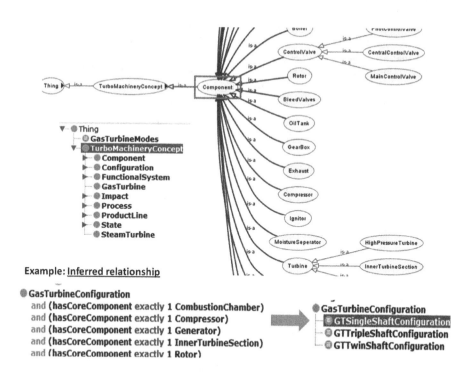

Fig. 5. Power generation - turbine ontology example

---

[1] https://www.w3.org/2005/Incubator/ssn/ssnx/ssn.

6   Challenges of the Semantic Approach

Although early semantic-based KM approaches have shown the benefits of using ontologies to support the KM life cycle, there still exist a large number of challenges from the automation and digitalization of industrial resources. These have to be addressed in order to make semantic technologies fully functional in an industrial setting. In this paper, we aim to answer the following questions:

- How to support analytics and make analytical workflows closer to the data? This new situation demands that semantic paradigm should be able to adjust according to the available data sources and make analytics easier to implement and deploy.
- How to remain 'abstract'? This means that semantic-based interfaces should not only support data-access and integration but also help end-users in developing semantic-based analytics and reporting workflows. Thus, an appropriate level of abstraction is required over existing data sources, analytical tools and reporting technologies.
- How to describe analytics outcome? A level of abstraction is required to define the outcomes from analytics and to use the results for decision-support tasks and reporting mechanisms.
- How to cater Big-Data? Semantic-based solution must adhere to the requirements of big-data architectures, where semantic layer can be adopted to manage, reuse and share large-scale data.
- How to manage authoring problems? Industries that have shallow standards and few specialized tools encounter problems with domain modelling and efficient authoring. Thus, semantic solution should be able to support use of reporting mechanism, workflows and design templates to understand, find and display relevant information.

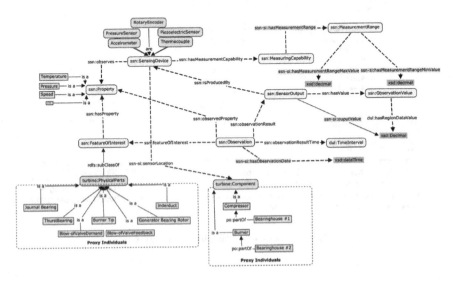

Fig. 6. Extended turbine ontology

7   Our Proposed Solution

Figure 7 shows the architecture of our proposed solution. It comprises of several components. The Ontology-base data-access (OBDA) middleware provides an abstraction over the existing data sources including interfaces to very large data sets such as hadoop clusters or Spark. The OBDA middleware allow users to formulate their information requirement (i.e. queries) without any knowledge about the data model or source and retrieve the relevant data automatically. A set of mapping is maintained which describes the relationship between the terms in the ontology and the corresponding terminology in the data source specifications, e.g. table and column names in the relational database schemas.

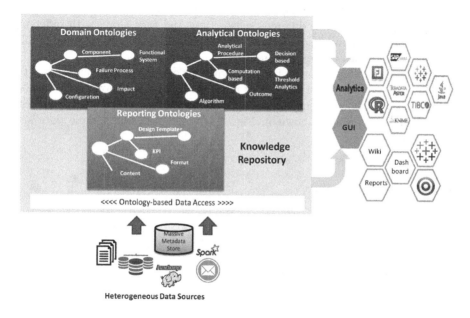

Fig. 7. New approach for semantic-based knowledge management

The domain ontology is extended to include concepts related to system configurations, failure processes, causality and more. Furthermore, the extension to analytics and reporting ontology is an important component of our solution. Knowledge about analytical workflows such as feature set or algorithms to be used and outcomes from analytics can be represented in terms of ontology to support automation and integration of knowledge resources for decision-support. Whereas, reporting ontologies represent different types of reports, content, design templates to be used and more. For example, *"Give me an analysis report X for all machines of type Y that had a'shut-down' in last three months"*.

Our solution also supports existing analytical and reporting workflows by exposing a Java-based OBDA node with SPARQL endpoint. This sort of integration can be made into any analytics environment (such as in KNIME or R analytics) and leverage the existing analytical models with semantic interfaces.

7.1 Preliminary Results from Siemens Turbo-Machinery Use-Case

We setup our solution for Siemens Turbo-machinery use-case where users analyse key performances indicators (KPI) of different gas turbines and compare the reliability, availability and outages as per different parameters (such as product, service region etc.). The difficulty today is that various machines have different sensors that contribute towards performance, their location within the partonomy may also vary and sensor tags are unknown. Today, users enumerate all sensor tags by hand and formulate customized KPI rules for each machine individually because they have different sensor type and tags along with different threshold values. With our solution, we overcome these problems by using ontology, OBDA mappings and exposing all this to analytical tool.

Figure 6 shows SSN ontology together with our extended turbine ontology to exemplify the idea of capturing part of relations between unknown individuals, sensors, measurable, and sensor meta-data including measurement capabilities. Use of OBDA mappings (shown in Fig. 8) is to connect the sample data set against ontology.

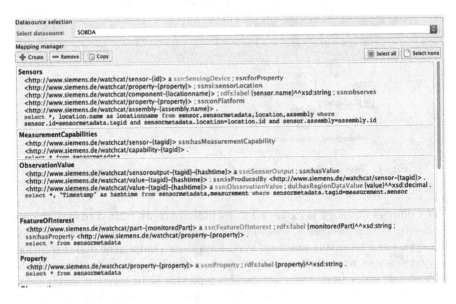

Fig. 8. Semantic mappings

As users were well-equipped with KNIME analytics tool, we provided an integration of our semantic node into this analytics platform. The added-value

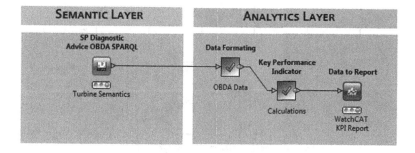

Fig. 9. Semantic node for KNIME-based workflow

Fig. 10. SPARQL query to access turbine data

of our solution is the automation of KPI calculations across machines by using single KPI analytical workflow to determine service hours, period hours and outages.

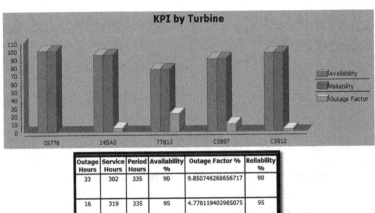

Fig. 11. Results from KPI report

Figure 9 shows a KNIME[2] based workflow where the our implemented source node for turbine semantics is exposed to computational procedure of performance analysis.

Figure 10 shows a detailed view of the SPARQL query that uses terminologies from the domain ontology to extract data and resulting reports are made available to the user. Figure 11 shows snapshots of such visualizations.

8   Conclusion

In order to make effective use of knowledge, it needs to be classified, defined and related in conventional terminologies. In this paper, we have discussed the basic requirements and limitations of knowledge management systems along with existing semantic and non-semantic approaches to knowledge acquisition, representation, modelling, discovery and distribution. We have presented results from the life sciences and engineering where ontology is used for various application tasks, together with an analysis of the feature set that semantic reasoning brings to the domain. Furthermore, we have described the current challenges for semantic approach and proposed a new solution to solve for big data-access, analytics, reporting and automation. Preliminary results have been presented for Siemens Turbo-machinery use-case where we used the OWL-QL language to define components of turbine ontology, and identified axioms involved in entities - and their

---

[2] https://www.knime.org/.

interactions to analyse turbine performance indicators. The semantic node was integrated into analytics platform KNIME to automate the workflow and make knowledge discovery tasks easy to follow.

References

1. Jones, J.L.: Robots at the tipping point: the road to iRobot Roomba. IEEE Robot. Autom. Mag. **13**(1), 76–78 (2006)
2. Carroll, J.M., Rosson, M.B.: Getting around the task-artifact cycle: how to make claims and design by scenario. ACM Trans. Inf. Syst. (TOIS) **10**(2), 181–212 (1992)
3. Chen, H., Fuller, S.S., Friedman, C., Hersh, W. (eds.): Medical Informatics: Knowledge Management and Data Mining in Biomedicine, vol. 8. Springer, Berlin (2006). https://doi.org/10.1007/b135955
4. Sajeva, S.: Critical analysis of knowledge management maturity models and their components. Econ. Manage. **14**, 611–623 (2015)
5. Edwards, J.S.: Knowledge management concepts and models. In: Bolisani, E., Handzic, M. (eds.) Advances in Knowledge Management. KMOL, vol. 1, pp. 25–44. Springer, Cham (2015). https://doi.org/10.1007/978-3-319-09501-1_2
6. Smith, B., Welty, C.: Ontology: towards a new synthesis. In: Formal Ontology in Information Systems, vol. 10, no. 3, pp. 3–9. ACM Press, USA, pp. iii-x, October 2001
7. Liebowitz, J.: Knowledge management and its link to artificial intelligence. Expert Syst. Appl. **20**(1), 1–6 (2001)
8. Russell, S., Norvig, P., Intelligence, A.: A Modern Approach. Artificial Intelligence. Prentice-Hall, Egnlewood Cliffs (1995). 25, 27. Chicago
9. Delone, W.H., McLean, E.R.: The DeLone and McLean model of information systems success: a ten-year update. J. Manage. Inf. Syst. **19**(4), 9–30 (2003)
10. Alavi, M., Leidner, D.E.: Review: knowledge management and knowledge management systems: conceptual foundations and research issues. MIS Q. **25**, 107–136 (2001)
11. Davenport, T.H., De Long, D.W., Beers, M.C.: Successful knowledge management projects. Sloan Manage. Rev. **39**(2), 43 (1998)
12. Chircu, A.M., Kauffman, R.J.: Limits to value in electronic commerce-related IT investments. J. Manage. Inf. Syst. **17**(2), 59–80 (2000)
13. Damodaran, L., Olphert, W.: Barriers and facilitators to the use of knowledge management systems. Behav. Inf. Technol. **19**(6), 405–413 (2000)
14. Evgeniou, T., Cartwright, P.: Barriers to information management. Eur. Manage. J. **23**(3), 293–299 (2005). Chicago
15. Benbya, H., Passiante, G., Belbaly, N.A.: Corporate portal: a tool for knowledge management synchronization. Int. J. Inf. Manage. **24**(3), 201–220 (2004)
16. Joo, J., Lee, S.M.: Adoption of the Semantic Web for overcoming technical limitations of knowledge management systems. Expert Syst. Appl. **36**(3), 7318–7327 (2009)
17. Davies, J., Fensel, D., Van Harmelen, F. (eds.): Towards the Semantic Web: Ontology-driven Knowledge Management. Wiley, New York (2003)
18. Donnelly, K.: SNOMED-CT: the advanced terminology and coding system for eHealth. Stud. Health Technol. Inform. **121**, 279 (2006)

19. Kharlamov, E., et al.: Optique: towards OBDA systems for industry. In: Cimiano, P., Fernández, M., Lopez, V., Schlobach, S., Völker, J. (eds.) ESWC 2013. LNCS, vol. 7955, pp. 125–140. Springer, Heidelberg (2013). https://doi.org/10.1007/978-3-642-41242-4_11
20. Mehdi, G., Brandt, S., Roshchin, M., Runkler, T.: Semantic Framework for Industrial Analytics and Diagnostics
21. Kharlamov, E., et al.: Capturing industrial information models with ontologies and constraints. In: Groth, P., et al. (eds.) ISWC 2016. LNCS, vol. 9982, pp. 325–343. Springer, Cham (2016). https://doi.org/10.1007/978-3-319-46547-0_30
22. Uschold, M., Gruninger, M.: Ontologies: principles, methods and applications. Knowl. Eng. Rev. 11(02), 93–136 (1996)

Author Index

Printed in the United States
By Bookmasters